GUIDE
TO THE
UNIVERSITY BOTANIC GARDEN
CAMBRIDGE

FRONTISPIECE

THE PTEROCARYA THICKET IN WINTER (NORTH-WEST
CORNER OF GARDEN) (p. 37)

GUIDE
TO THE
UNIVERSITY BOTANIC GARDEN
CAMBRIDGE

BY

HUMPHREY GILBERT-CARTER
DIRECTOR OF THE GARDEN

CAMBRIDGE
AT THE UNIVERSITY PRESS
1922

CAMBRIDGE UNIVERSITY PRESS
Cambridge, New York, Melbourne, Madrid, Cape Town,
Singapore, São Paulo, Delhi, Mexico City

Cambridge University Press
The Edinburgh Building, Cambridge CB2 8RU, UK

Published in the United States of America by Cambridge University Press, New York

www.cambridge.org
Information on this title: www.cambridge.org/9781107643178

© Cambridge University Press 1922

This publication is in copyright. Subject to statutory exception
and to the provisions of relevant collective licensing agreements,
no reproduction of any part may take place without the written
permission of Cambridge University Press.

First published 1922
First paperback edition 2013

A catalogue record for this publication is available from the British Library

ISBN 978-1-107-64317-8 Paperback

Cambridge University Press has no responsibility for the persistence or
accuracy of URLs for external or third-party internet websites referred to in
this publication, and does not guarantee that any content on such websites is,
or will remain, accurate or appropriate.

INTRODUCTION

THE sequence of the Families and Genera in this *Guide* is that of the eighth edition of the *Engler-Gilg Syllabus der Pflanzenfamilien* published in 1919. Orders are not mentioned.

It may be necessary to explain that the word *Family* in modern use is equivalent to the *Natural Order* of the older English Botanists, and that the modern *Order* is equivalent to the old-fashioned *Cohort*. The International Rules of Botanical Nomenclature adopted by the International Botanical Congress of Vienna in 1905, and Brussels, 1910, recommend that the words be used in this way; and to use them otherwise causes difficulty and confusion. The same rules recommend that the names of Orders should end in *-ales* and those of Families in *-aceae*, but they allow other terminations to be retained provided that they do not lead to confusion or error. For example the following names of Families from long usage are exceptions to the rule: *Palmae, Gramineae, Cruciferae, Leguminosae, Guttiferae, Umbelliferae, Labiatae, Compositae*.

The name of each species consists of two words. The first word is the *generic* name, or name of the *genus* to which the species belongs, the second is the *trivial* name. The two together constitute the *specific* name. Generic names are always written with capital letters. Certain trivial names, too, are usually written by botanists with capital letters, but in this book the practice of zoologists, and of Moss in the *Cambridge British Flora* has been followed by writing all trivial names with small letters. The name of the author of the name is placed, usually in an abbreviated form, after the trivial name. This is necessary because different botanists have sometimes given the same name to different plants. When this has happened, according to the international rules, the plant first so-named is allowed to keep the name.

INTRODUCTION

For example, the Wych Elm in this country is well known by the name *Ulmus montana*. This name was given by Stokes in 1787, and is therefore written *Ulmus montana* Stokes. As the Wych Elm had already been named *U. glabra* by Hudson in 1762, the name *Ulmus montana* is not valid for it. The proper name for the Wych Elm is *Ulmus glabra* Hudson. Again, the name *Ulmus glabra* is often applied to the Smooth-leaved Elm, to which it was given by Miller in 1768. As this name had already been given to the Wych Elm in 1762, it cannot be used for the Smooth-leaved Elm, the proper name of which is *Ulmus nitens* Moench.

From these remarks it will be seen that the Law of Priority rules the questions of Botanical Nomenclature, and that the Law of Aptness holds no sway. This may seem unfair legislation, but in practice it is the only legislation possible. It is usually easy to decide which is the prior of several names, whereas the comparative aptness of names is often a matter of opinion. The starting point of the Nomenclature of the Flowering Plants and Ferns is the *Species Plantarum* of Linnaeus published in 1753.

Selecting the materials for the *Guide* has been a difficult task. The Botanic Garden of a University is the *alma ancilla* of that University's Botany School, whose refulgence it should reflect. But certain subjects are obviously better suited than others for mention in a *Guide*. For example Ecology must figure more largely than Fossil Botany, and Plant Physiology and Mycology can scarcely occupy any space at all.

It has been the happy destiny of this garden to be loved and befriended by nearly all the notable oriental scholars of the University. This connection between Oriental Studies and Botany has prompted the author to give certain eastern names of plants and quotations illustrating the use of these names.

The various Indian names mentioned are those commonly used in the north of India by Hindustani-speaking Indians. For most of the Indian words I have chosen the Arabic alphabet rather than the Nagari because it is the

INTRODUCTION

proper alphabet of the dialect of Hindustani called Urdu, which, chiefly because we are very susceptible to its beauties, is much better known to the English than Hindi, whose proper alphabet is the Nagari or Sanskrit.

The author is very grateful for the assistance received during the preparation of this *Guide*. He is particularly indebted to Professor A. C. Seward, for help with the Gymnosperms; to Mr A. G. Tansley, for much help with the ecological portions; to Mr F. J. H. Jenkinson, for kindly allowing him to use the University Library, and to the staff of the University Library, particularly to Mr E. J. Thomas, for his sound guidance along many paths; to Professor E. G. Browne, for help with the Arabic and Persian, and for kindly allowing him to use his translations of some of the hemistichs quoted; to Professor H. A. Giles for help with Chinese words; to Professor Rapson for help with Sanskrit words; to Dr F. H. H. Guillemard for reading the proofs, and for several valuable suggestions; to Mr F. G. Preston, the Superintendent of the Botanic Garden, for many valuable suggestions, and for taking the photographs for several of the plates; to the editors of the *Garden*, the *Gardener's Chronicle*, and *Country Life* for permission to use plates published in their papers; to the Royal Horticultural Society for permission to reproduce a plate from the *Journal* of the Society; and to Mr Debenham and his pupils for making the map from which the plan of the garden has been prepared.

It is with special pleasure that I express my thanks to a friend to whom the University is indebted for more than one benefaction to the Botanic Garden. It is through his generosity that the publication of this *Guide* has been made possible.

H. G-C.

August 1922

HISTORICAL NOTE

WE read in the *Cambridge Portfolio* that as long ago as 1696 the ground for a Physic Garden had been measured and the plan drawn, but through some unknown impediment the scheme failed. In 1724 Professor Bradley made large but hollow promises on the subject which he publicly repeated in his lectures in 1729; but nothing was done. In 1731 there appeared more hope; for many conferences were held between the Vice-Chancellor, Professor John Martyn, and Mr Philip Miller, of the Chelsea Garden, respecting the estate of a Mr Brownell of Willingham, which was once intended to be devoted to the establishment of a Botanic Garden at Cambridge; but this estate was diverted into another channel. At length the plan was happily effected through the liberality of Dr Walker, the Vice-Master of Trinity College, who gave an estate to trustees for that purpose. The ground selected was the site of the Monastery of the Austin Friars, in the parish of St Edward's, and was purchased by Dr Walker for £1600 in 1761.

The site of this old Garden, together with five or six tenements in Free School Lane, amounting in all to over five acres, were made over to the University by an Indenture dated 24th Aug. 1762, which is in the Registrary's office.

In 1831 an Act of Parliament was obtained authorising the removal of the old Garden to its present site. The iron gates which now guard the main entrance (Pl. I) were removed from the old Garden in 1909.

Laboratories now stand on the site of the old Garden, whose sole relic is a magnificent specimen of the Chinese tree *Sophora japonica* (Pl. II) which stands beside the Pathological Laboratory.

Further historical information about the Garden will be found in the *University Historical Register* (p. 214), and in the *Cambridge Portfolio* (p. 81).

CONTENTS

	PAGE
INTRODUCTION	v
HISTORICAL NOTE	viii
LIST OF ILLUSTRATIONS	xii
GLOSSARY	xiii
NOTE ON LEAVES	xiv
BIBLIOGRAPHY	xvi

GYMNOSPERMAE

CYCADACEAE	1
GINKGOACEAE	2
TAXACEAE	3
PINACEAE	4
GNETACEAE	12

ANGIOSPERMAE
MONOCOTYLEDONEAE

GRAMINEAE	15
CYPERACEAE	16
PALMAE	17
ARACEAE	19
BROMELIACEAE	23
LILIACEAE	25
AMARYLLIDACEAE	28
MUSACEAE	29
ZINGIBERACEAE	31

CONTENTS

DICOTYLEDONEAE

ARCHICHLAMYDEAE

	PAGE
CASUARINACEAE	32
SALICACEAE	32
MYRICACEAE	37
JUGLANDACEAE	37
BETULACEAE	39
FAGACEAE	43
ULMACEAE	45
MORACEAE	47
NYMPHAEACEAE	51
BERBERIDACEAE	51
MENISPERMACEAE	53
MAGNOLIACEAE	53
ANONACEAE	55
LAURACEAE	56
CAPPARIDACEAE	57
SARRACENIACEAE	57
NEPENTHACEAE	58
DROSERACEAE	58
CEPHALOTACEAE	59
SAXIFRAGACEAE	60
HAMAMELIDACEAE	60
EUCOMMIACEAE	62
PLATANACEAE	62
ROSACEAE	63
LEGUMINOSAE	65
ERYTHROXYLACEAE	73
RUTACEAE	73
SIMARUBACEAE	75
BUXACEAE	76
CORIARIACEAE	77

CONTENTS

	PAGE
ANACARDIACEAE	77
CELASTRACEAE	79
ACERACEAE	80
HIPPOCASTANACEAE	82
SAPINDACEAE	82
RHAMNACEAE	83
ELAEOCARPACEAE	85
TILIACEAE	85
MALVACEAE	87
STERCULIACEAE	88
CAMELLIACEAE	89
GUTTIFERAE	90
CISTACEAE	91
FLACOURTIACEAE	91
CACTACEAE	92
MYRTACEAE	93
CORNACEAE	94

METACHLAMYDEAE

ERICACEAE	96
EPACRIDACEAE	97
EBENACEAE	97
OLEACEAE	98
APOCYNACEAE	100
ASCLEPIADACEAE	100
VERBENACEAE	101
SOLANACEAE	101
SCROPHULARIACEAE	103
RUBIACEAE	103
CUCURBITACEAE	104
COMPOSITAE	105
INDEX	107

LIST OF ILLUSTRATIONS

The Pterocarya Thicket in Winter (north-west corner of Garden) . . . . FRONTISPIECE

PLATE
I. The Main Entrance and the Gates of the Old Botanic Garden . . . . TO FACE P. 4
II. The huge specimen of *Sophora japonica* on the Site of the Old Botanic Garden 5
III. The Neosia (*Pinus gerardiana*), near the Rock Garden 12
IV. The One-Needled Pine (*Pinus monophylla*), in the Pinetum 13
V. *Monstera deliciosa*, in the Palm House . . . 20
VI. The Golden-Club (*Orontium aquaticum*), in the Water Garden 21
VII. A mass of Red-Hot Poker (*Kniphofia caulescens*), at the entrance of Bay No. 4 28
VIII. The great Fan Aloe (*Aloë plicatilis*) on its way to its present home 29
IX. *Bomarea cantabrigiensis*, in Bay No. 4 . . . 36
X. Bole of the Paper Birch (*Betula papyrifera*) . . 37
XI. The Oak beside the Broad Walk 44
XII. *Asimina triloba*, flowering spray 45
XIII. *Asimina triloba* in flower, between the Stream and the Pond 52
XIV. An Australian Sundew (*Drosera binata*), in the Tropical Aquarium 53
XV. The Musk Rose (*Rosa moschata*), climbing up the Austrian Pines on the Hill 60
XVI. The Judas Tree (*Cercis siliquastrum*) 61
XVII. The Hedgehog Broom (*Erinacea pungens*), in Bay No. 2 68
XVIII. The Trifoliate Orange (*Aegle sepiaria*), between Bays Nos. 5 and 6 69
XIX. The Tree of Heaven (*Ailanthus glandulosa*), near the Stream 76
XX. *Xanthoceras sorbifolia*, beside the Superintendent's House 77
XXI. *Xanthoceras sorbifolia*, a Flowering Spray . . 84
XXII. *Clusia grandiflora*, in the Stove 85
XXIII. Opuntias in the open by the Lynch Walk . . 92

Plan of the Garden
available for download from www.cambridge.org/9781107643178

GLOSSARY

Androecium, the stamens of a flower taken collectively.

Apocarpous, with the carpels free one from another.

Berry, a fleshy fruit whose seeds lie free in the pulp (*see* Drupe).

Coriaceous, leathery.

Deciduous (of trees), shedding the leaves annually.

Dioecious, having staminate ('male') and carpellary ('female') flowers on separate plants.

Drupe, a fleshy fruit whose seed is contained in a hard stone (e.g. cherry, plum, olive).

Epiphyte, a plant that lives, unattached to the soil, upon another plant without being parasitic upon it (*see* Parasite).

Fastigiate, having many branches which all ascend parallel to the main stem (e.g. Lombardy Poplar).

Glaucous, sea-green.

Parasite, a plant that draws some or all of its food from another plant.

Scape, a leafless or nearly leafless flower-bearing stem which arises from the base of the plant.

Sucker, a shoot springing from the underground parts of trees and shrubs.

Syncarpous, having the carpels united together.

Whorl, several leaves arranged around the stem at one level.

Xerophilous, adapted to live where the water supply is limited.

NOTE ON LEAVES

LEAVES must not be looked upon merely as the organs which, by means of the sun's rays, build up simpler substances into the more complicated substances which are indispensable to life, but it also must be considered that it is through its leaves that the plant loses water. In general the larger the leaf the greater the loss of water a plant must suffer. Thus it becomes of interest to contemplate the leaves in relation to the plant's supply and loss of water. Plants that live in parts of the world where there is a dry season of sufficient length to embarrass water supply, lose their leaves during that season. Most of our British trees lose their leaves in winter, and our perennial herbs lose not only their leaves but also the shoots that bear them. This is not so much because cold in itself is necessarily injurious to leaves as because cold roots cannot absorb water. To the plant a dry season and a cold season are alike periods of drought Evergreen plants prevail in regions where neither drought nor cold is of sufficient severity or continuance to compel leaf-fall. The following are two interesting types of evergreen leaf.

Sclerophyllous type. Leaves of this type are small, usually entire, and rather thick and rigid.

Vegetation whose characteristic component plants bear them is peculiar to regions where the summers are hot and dry, and rain falls during the winters, which are mild.

Of sclerophyllous plants we have in the Garden, from the Mediterranean Region, for example, the Holm-oak (*Quercus ilex* L., Family *Fagaceae*), and the Olive (*Olea europaea* L., Family *Oleaceae*); from California, *Castanopsis chrysophylla* DC. (Family *Fagaceae*), and species of *Ceanothus* (Family *Rhamnaceae*). In the Temperate House will be found many sclerophyllous plants from the Australian scrub, including *Styphelia richei* Labill. (Family *Epacridaceae*), species of *Melaleuca*, *Leptospermum*, and *Callistemon* (all belonging to the Family *Myrtaceae*), and various *Proteaceae*. In the same house will be found sclerophyllous plants from South-west

NOTE ON LEAVES

Africa such as *Gnidia polystachya* Berg. (Family *Thymelaeaceae*) and species of *Phylica* (Family *Rhamnaceae*). From Chile we have *Azara microphylla* Hook. f. (Family *Flacourtiaceae*) and species of *Escallonia* (Family *Saxifragaceae*). Many other examples will be found in the Garden.

The sclerophyll is well adapted to these regions, where on the one hand neither the dry heat of summer nor the cold of winter is sufficient to compel the plants to abandon their leaves, and on the other hand the risk of drought is too great to permit the existence of a large leaf surface. None of our British plants have leaves of this type.

Laurel type. This is allied to the sclerophyll, and like it is evergreen, hairless, entire, and generally ovate or elliptical, but it differs from the sclerophyll in being much larger, less hard and rigid, and in having a surface which shines by the reflection of light. The regions it affects are more humid than those whose characteristic vegetation is sclerophyllous. Plants with leaves of this type are often called 'Laurels.'

Familiar examples are *Prunus lusitanicus* (Portugal Laurel), *Prunus laurocerasus* L., the Cherry Laurel (Family *Rosaceae*), and *Aucuba japonica* Thunb. (Family *Cornaceae*), which do not belong to the genus *Laurus*, or even to the family *Lauraceae*, yet are called laurels because their leaves are of the laurel type.

Examples of British plants whose leaves are of this type are the Holly (*Ilex aquifolium* L., Family *Aquifoliaceae*) and the Ivy (*Hedera helix* L., Family *Araliaceae*).

The leaves of some plants are intermediate between the sclerophyllous and laurel types. Familiar examples are the Strawberry Tree (*Arbutus unedo* L., Family *Ericaceae*), and the Laurustinus (*Viburnum tinus* L., Family *Caprifoliaceae*).

* * * * * *

An attempt has been made in this *Guide* to enable readers to identify our several species of Birch, Oak, Elm, and Lime, by short descriptions of the leaves of these trees. For this purpose the leaves on the older boughs of the trees should be examined, as those on suckers and young adventitious shoots often differ widely from the mature type.

BIBLIOGRAPHY

BEAN, W. J. Trees and Shrubs hardy in the British Isles. 3rd edition. London, 1921.

CHAMBERLAIN, J. C. The Living Cycads. Chicago.

LYNCH, R. I. Trees of the Cambridge Botanic Garden, in the *Journal of the Royal Horticultural Society*, XLI. August, 1915.

MOSS, C. E. The Cambridge British Flora. Cambridge, II. 1914. III. 1921.

SARGENT, S. S. Manual of the Trees of North America. Boston, 1905.

TANSLEY, A. G. Types of British Vegetation. Cambridge, 1911.

TRISTRAM, H. B. The Natural History of the Bible. 10th edition. S.P.C.K. 1911.

WILLIS, J. C. A Dictionary of the Flowering Plants and Ferns. 4th edition. Cambridge, 1919.

GYMNOSPERMAE

Family CYCADACEAE

The 9 genera, containing 75 species, of living Cycads, all of which inhabit warm regions, are all that remain of a group of plants that was, perhaps, as characteristic of Mesozoic vegetation as Dicotyledons are of the flora of to-day. The stems of Cycads, which are often very short, and sometimes tuberous and subterranean, are surmounted by crowns of large, pinnate foliage. The stems of *Cycas* are clad in a mantle of alternating zones of thin scales and stout leaf-remnants. In the other genera the scales and leaf-remnants are more or less interspersed, and in some species of *Macrozamia* leaf-remnants alone are present.

A few Cycads are tall trees; *Macrozamia hopei*, native of northern Queensland, reaches a height of 60 feet. It will be noted that both tree-ferns and palms bear a superficial likeness to the Cycads.

The 'flowers,' which are always dioecious, and usually terminal, in nearly all Cycads resemble the familiar cones of the Coniferae. Our collection of *Cycadaceae* is in the Palm House.

The article 'Cycadaceae' in 'Willis' contains a key to the genera. This book and Chamberlain's *Living Cycads* are in the cupboard in the Corridor.

CYCAS

Cycas has 16 species, native of the Old World. Each leaflet has a conspicuous, unbranched midrib. The carpellary flower is not a cone, but a collection of leaf-like carpels through which the stem ultimately continues its growth. Most of the species of this genus have tall columnar stems. *C. media* R. Br. is the Nut Palm of Australia. It seldom

CYCADACEAE

exceeds 20 feet in height. See Chamberlain, *The Living Cycads*, p. 38.

C. revoluta *Thunb.* is native of southern Japan, and is planted in most hot countries. Its trivial name refers to the margins of the leaflets, which are rolled backwards.

STANGERIA

S. paradoxa *Moore*, the only species, is native of Natal. Its leaflets are pinnately veined. The leaflets of all other Cycads, except this plant and the species of *Cycas*, are traversed by numerous parallel nerves. A glance at the leaves dismisses surprise that the plant, before the flowers were known, was taken for a fern.

ENCEPHALARTOS

Encephalartos has about 20 species, natives of Africa. The leaves are often formidably armed. Kaffirs prepare a kind of meal from the pith of some species. The stems of *E. laurentianus* attain the height of over 30 feet.

DIOON

Dioon has 2 species, natives of Mexico.

D. edule *Lindl.* has curious flat rigid leaves. The seeds, which are about as big as chestnuts, are ground into meal.

The names of the other genera of the Cycadaceae are *Bowenia*, *Macrozamia*, *Zamia*, *Ceratozamia*, and *Microcycas*.

Family GINKGOACEAE

GINKGO

Ginkgo has only one living species. There is a small specimen on the bed opposite the Yew Hedge, and a large one beside the great cedar in the north-east corner of the Garden.

G. biloba *L.*, The Maidenhair Tree, so called because its fan-like leaves with forked veins resemble the ultimate divisions of the fronds of maidenhair ferns. This tree is no longer found wild. Its last stronghold was probably China, where it is held sacred and preserved in temple gardens.

GINKGOACEAE

Fossil leaves indistinguishable from those of the living tree are preserved in Tertiary sedimentary beds between the basalts of the Island of Mull. Species of *Ginkgo* occur also in the Jurassic rocks of the Yorkshire coast, and in other Mesozoic strata in nearly all parts of the world. There is probably no other existing tree which has so strong a claim to the title of a 'living fossil.' The Maidenhair Tree is dioecious, and most of the large specimens in the British Isles are staminate.

Family TAXACEAE

The two families *Taxaceae* and *Pinaceae* together constitute the class called *Coniferae*. There is a collection of hardy *Taxaceae* at the east end of the Houses.

PHYLLOCLADUS

Phyllocladus has 6 species, inhabiting Tasmania, New Zealand, and Borneo. The apparent leaves of these plants are really shoots. They arise in the axils of scale-leaves and may themselves bear scale-leaves on their edges. Leaf-like shoots are termed *cladodes* or *phylloclades* and are seen also in *Ruscus* (Butcher's Broom), *Semele*, and other genera. Their position relative to buds and scale-leaves should be compared with that of the *phyllodes* (modified petioles) of *Acacia*, section *Phyllodineae* (p. 65). Our specimens of *Phyllocladus* are in the Temperate House.

P. trichomanoides *D. Don*, Celery-leaved Pine. Native of New Zealand. The bark of this plant is used for tanning and yields a red dye. The stems make excellent walking sticks, and if the young growing stems are bruised, the dye escapes into the wood, giving the sticks a beautiful mottled appearance. Its native name *Tanekaha* is said to mean 'strong in growth.'

TAXUS (YEW)

Taxus has a single species, native of the North Temperate Region. The various geographical forms of the Yew are often considered to be distinct species. *Taxus* is the only genus of *Taxaceae*, and indeed of all the *Coniferae*, from which resin-canals are absent. A Yew hedge extends from the Bicycle Enclosure to the entrance of the Houses, and there is a collection of Yews in the angle between the Border Walk and Middle Walk.

T. baccata *L.*, Common Yew. Native of the North Temperate Region, including Britain. The foliage of the Yew superficially resembles that of the Firs (*Abies*), but the leaves are bright green beneath, and lack the two longitudinal stripes of the Firs. The ripe seed is surrounded by a bright red fleshy aril, which is edible. In the Middle Ages the Yew was of great importance because it furnished the best wood for bows. The tree attains a great age. It is more abundant and widespread in the British Isles than in most West European countries. There are pure native Yew woods on the southern chalk, and these cast a deeper shade than any other British tree, so that the ground beneath is quite bare. It is very commonly planted in churchyards. As a 'funereal' tree it corresponds with the Cypress (*Cupressus sempervirens* L., p. 10) of Mediterranean countries. The Irish Yew (var. *fastigiata*) is always a carpellary tree. It was discovered wild in Ireland in 1780. All existing trees are ultimately from cuttings from a tree in Florence Court in Co. Fermanagh. Besides being fastigiate this variety differs from the type in having its leaves standing out on all sides of the twigs, and not disposed in one plane. Cattle are sometimes poisoned by eating the foliage of the Yew.

Family PINACEAE

CEDRUS

Cedrus is usually stated to have 3 species, but these are very closely allied and are perhaps only geographical forms of one species. The dwarf shoots bear very numerous, persistent leaves.

C. libani *Barrelier*, The Cedar of Lebanon. Native of Mount Lebanon and of the Cilician Taurus in Asia Minor. This tree is often mentioned in the Old Testament, but it

THE MAIN ENTRANCE AND THE GATES OF THE OLD BOTANIC GARDEN (p. viii)

PLATE II

THE HUGE SPECIMEN OF *SOPHORA JAPONICA* ON THE SITE OF THE OLD BOTANIC GARDEN (p. viii)

PINACEAE

is possible that the Hebrew word אֶרֶז, which is translated 'Cedar,' is used in some passages to include certain other trees (compare Arabic ارز). Engler includes in this species the Atlantic Cedar (**C. atlantica** *Manetti*), which grows on the Atlas Mountains in Algeria. A large specimen with greyish-blue leaves stands beside the Cedar of Lebanon near the Rock Garden.

C. deodara *Lawson*, Deodar. Native of the Western Himalaya. The word 'Deodar' is the Indian vernacular देवदार which is a shortened form of देवदारु meaning 'Tree of the Gods.' In Bengal this name, and a variant देबदार, are applied to *Polyalthia longifolia* Benth. et Hk. (Fam. *Anonaceae*). Both these trees are held in reverence by the Hindus. The wood of the Deodar is a well-known medicine in India. It is called भद्रकाष्ठ (auspicious wood), and is usually administered in powder combined with other drugs.

Pinus *L.* (Pine)

Pinus has about 80 species, for the most part native of the North Temperate Region. The short shoots have usually 2, 3, or 5 leaves (needles). Most of our specimens will be found in the Pinetum, but some are by the Main Walk, and in other parts of the Garden.

P. excelsa *Wallich*, The Himalayan Blue Pine. Native of the Himalaya, where it grows to a height of 150 feet. It was introduced into this country in 1823. The cones of the section of five-needled Pines called *Strobus* are at least three times as long as broad, and their scales differ widely from those of the familiar Scotch Pine and Austrian Pine, which belong to the section *Sylvestres*. The needles of this species are bent abruptly near the base in such a manner that the greater part of the needle hangs downwards.

The more resinous parts of the wood of this tree make splendid torches. In its native land during dry winters there sometimes appears on the leaves and twigs a sweet, manna-like substance which the natives collect and eat.

PINACEAE

P. gerardiana *Wall.*, The Neosia or Edible Pine (Pl. III). A three-needled Pine, native of the arid valleys of the N.W. Himalaya. The almond-like seeds (called in India چلغوزه) are edible, and thought by some Indians to have wonderful medicinal qualities. These seeds are the daily bread of the people of Kunáwár who say that one tree is a man's life in winter. Considerable quantities are exported to the plains of India.

The bark of this tree peels off in large scales like that of the Planes (*Platanus*). Our specimen is probably the finest in the country (see Lynch, p. 2). It is now 22 feet 10 inches high.

The timber is hard, durable, and very resinous, but because of the value of the seeds, the trees are seldom felled.

P. monophylla *Torrey*, One-needled Pine (Pl. IV). Native of Utah, Nevada, Arizona, and Lower California. In some parts it forms extensive open forests at altitudes of between five and seven thousand feet. The leaves of this Pine, though usually solitary, are sometimes in pairs. The seeds are eaten by the Indians of Nevada and California (see Lynch, p. 7). Our specimen is now 11 feet 10 inches high.

P. pinaster *Solander*, Maritime Pine, Cluster Pine. Native of South Europe from western France to Greece. The Maritime Pine forms pure woods on the great masses of siliceous rock which compose the mountain groups of the Maures and Estérel, but the characteristic Pine of the limestone rocks of that region is *P. halepensis* Miller, the Aleppo Pine. The very long, stiff, sharp-pointed needles are arranged in pairs. The large cones have a rich brown 'lacquered' surface. This is the commonest tree in the Pine woods of Bournemouth. Though its timber is of little use, its resin is very valuable. "Nowhere has its economic value been so efficiently demonstrated as in the Landes of France, south of Bordeaux. Here in 1904, mostly planted by man, it covered an area of about 1¾ million acres, yielding an annual revenue of £560,000, and from land which previously was mainly desert" (Bean, II, 188). The Germans often call this tree 'Sternkiefer,' but the suffix 'aster' denotes merely a wild tree, and is not connected with the Greek ἀστήρ, star. Pliny

PINACEAE

used the word *Pinus* for a tree that grew in gardens, and *Pinaster* for one that grew wild.

P. sylvestris *L.*, Scotch Pine. The range of this, our only British species, is wider than that of any other Pine. It is native of nearly all parts of Europe and extends across Siberia to the region of the Amur River. It grows wild in parts of the Highlands of Scotland. (The Scotch form is sometimes called var. *scotica*.) In England seedlings from plantations have colonised large areas of the southern sands.

P. laricio *Poiret*, var. **nigricans** *Parlatore*, Austrian Pine. Some specimens of this very commonly planted tree stand on the Hill. It resembles the Scotch Pine, but has neither its rich red, scaling bark, nor its beautiful, glaucous foliage.

In the Pinetum will be seen several large specimens of *P. laricio* var. *pallasiana*, which is probably native of the Crimea. In this variety the larger branches grow straight upwards. The true *P. laricio* (Corsican Pine) is a stately tree with paler, softer foliage[1].

SCIADOPITYS

S. verticillata *Sieb. et Zucc.*, The Umbrella Pine, is the only species. It is native of Japan, where it is planted in temple gardens. The two leaves of each short shoot are fused together to form a single needle. Examine these needles and note the longitudinal groove indicating the line of union of the two leaves. Our specimen is on a peat-bed on the east side of the West Walk.

SEQUOIA

Though in Cretaceous and Tertiary times species of this genus were widely scattered over the northern hemisphere, and their fossil remains have been found in England (at Bovey Tracey and elsewhere), both the species which are now living are confined to the mountains of California.

S. sempervirens *Endlicher*, The Redwood. This species has flat leaves which spread in two ranks. Sargent (p. 69) says that it may reach a height of 340 feet and that the

---

[1] Compare the Austrian Pine and Corsican Pine which stand together by the lawn near the south end of the Pinetum.

diameter at the base may be 28 feet. The timber is much valued for construction. The wood employed in the building of some Californian cities is almost entirely of this tree. The Redwood was introduced into this country about the middle of last century. We have a specimen by the Bicycle Enclosure, and one near the entrance to the Water Garden. This species produces suckers freely.

S. gigantea *Decaisne*, Wellingtonia, Big Tree. This species has awl-shaped leaves which are triangular in section, and arranged all round the branchlets. Though surpassed in height by *S. sempervirens*, and by the Eucalyptus trees of Australia, it is certainly in bulk the largest tree in the world. It occasionally reaches a height of 320 feet, with a diameter at the base of 35 feet. The Wellingtonia was introduced into this country by Mr W. Lobb in 1853. The timber is light, soft, and brittle. There are several specimens by the Main Walk. This species never produces suckers.

Taxodium

Taxodium has 3 or 4 species, native of North America. The Swamp Cypress grows on a bed between the West Walk and the Pond. The species of this genus have threefold interest. Firstly, they are among the few deciduous *Pinaceae*. Secondly, it is not merely the leaves that fall, but shoots of limited growth whose numerous leaves are arranged in two ranks. These shoots superficially resemble pinnate leaves. Thirdly, the roots of the Swamp Cypresses bear upright branches which project into the air as do the pneumatophores of Mangroves. These 'knees' develop only in swampy soil. Their function is probably to convey air to the water-logged roots. The ground around our specimen has long been converted into a swamp, but no knee has yet appeared.

T. distichum *Rich.*, Deciduous Cypress, Swamp Cypress, is native of the southern United States.

There is a picture of a huge specimen of a *Taxodium* in the Botany School Museum.

PINACEAE

Tetraclinis

Tetraclinis has 1 species. Its nearest allies grow in the southern hemisphere. Our specimens are in the Temperate House.

T. articulata *Masters* (*Callitris quadrivalvis* Vent.), Thyine Wood of Rev. xviii, 12. Native of the mountains of North Africa, of South-east Spain, and of Malta. Greek and Roman writers make frequent mention of the hard, fragrant wood of this tree. It is the θύον of Theophrastus (*Hist. Plant.* v, iii, 7), who knew that the tree resembled a Cypress, and who tells us that men could still recall that some of the roofs in 'ancient times' were made of its timber, which was proof against decay.

The Romans called it *Citrus* (cf. Greek κέδρος, Cedar, or other fragrant-wooded coniferous trees). They used the wood for protecting clothes against moths, but for a long time the common people knew nothing of the tree itself, and thought that Citrons, which were brought to Rome, and also used against moths, were the fruits of the tree which yielded Citrus wood. Hence they called this fruit Citrus, a name which it keeps to this day (*Citrus medica* L.). The wood is once mentioned in the Bible under the name 'Thyine wood' (Ξύλον θύϊνον). The powdered resin of this tree called 'Sandarach' was used as pounce to prevent ink from spreading on parchment.

Thuya (Arbor-vitae)

Thuya has 6 species, native of North America and Eastern Asia. The cones are composed of flat, overlapping scales.

T. occidentalis *L.*, Arbor-vitae. Native of Atlantic North America from Nova Scotia to Virginia, where it often forms nearly impenetrable forests on swampy or moist ground. From Gerard's Herbal we learn that this tree and *Yucca gloriosa* were cultivated in England in 1596. These were probably the first two woody American plants to be grown in this country. The durable, fragrant wood is used in Canada and the Northern States for fence-posts, etc. In this species the surface of the lower leaves bears a conspicuous gland, and the cones have usually four fully developed scales. There are some specimens south of the lawn.

T. plicata *D. Don*. Native of Pacific North America. In Britain this species grows more vigorously than the Common Arbor-vitae, which it is supplanting in our gardens. Foresters too are planting it as a timber tree. The foliage is whitish beneath, and the leaf-glands are inconspicuous. The cones have usually six fully developed scales. The hedge that runs between the Superintendent's House and the Border is of this species.

T. orientalis *L.* (*Biota orientalis* Endlicher), Chinese Arbor-vitae. Native of China, Japan, and Formosa. The frond-like branches of this species spread vertically; in other species of *Thuya* they are horizontal. The scales of the cones are horned, and the seeds are wingless. Our best specimens are south of the lawn.

CUPRESSUS (CYPRESS)

Cupressus has about 12 species, natives of the Mediterranean Region, Asia, and North America. The cones are globular and composed of shield-like scales which fit together edge to edge.

C. sempervirens *L.*, Italian Cypress. Native from Persia to South-east Europe. It has been planted in Italy from classical times, and in England for at least four centuries. There is a form with horizontal branches, but the fastigiate variety is the well-known tree characteristic of Italian landscapes, and the سرو of Persian poets, who liken the figure of youth to its graceful form:

قد موزونش همپایهٔ سرو کشمر-- قاآنی.

"*Her graceful form was like to Kashmar's Cypress fine.*"

The famous Cypress of Kashmar, said to have been planted by Zoroaster, was venerated by his followers. The Italian Cypress is often planted in both Christian and Muhammadan cemeteries. It looks as much like a closed umbrella as

PINACEAE

Pinus pinea L., the Stone Pine, resembles an open one. The timber is remarkably durable. As its aroma is agreeable to man and repellent to moths, it makes excellent clothes-chests.

C. lawsoniana *Murray.* Native of Pacific North America where it grows usually in small groves, but to the south of Cape Gregory it forms a nearly continuous forest-belt twenty miles long. The wood has many uses; on the Pacific Coast matches are made of it. Many varieties are in cultivation in Britain. Bean says that it is perhaps the commonest Conifer in our gardens.

JUNIPERUS (JUNIPER)

Juniperus has 30 species, mostly native of the North Temperate Regions. The scales of the cone while ripening become fleshy and fuse together forming a fruit like a berry. Most of our specimens are in the Pinetum.

The Juniper of the Old Testament is doubtless *Retama raetam* Coebb et Berth., a leguminous shrub. The Hebrew is רֹתֶם; the Arabs still call this plant رَتَم. See Tristram, p. 359, and Lane's *Arabic Lexicon*, Art. رتم.

J. communis *L.,* Common Juniper. Native of the North Temperate and Arctic Regions. It is scattered throughout the British Isles, and grows abundantly in the Highlands of Scotland. Note how the sharp-pointed leaves spread at right angles to the stem that bears them. The glaucous, blue-black fruits are used for flavouring gin. In Persia it is called عَرْعَر (see below), and جَرْجَرِي.

J. sabina *L.,* Savin. A shrub. Native of Central and South Europe, the Caucasus and North Asia. Note the difference in the shape and arrangement between the juvenile and adult leaves. The ripe fruits are borne on short stalks, and are bent backwards. This plant is probably the עַרְעָר of Jer. xvii, 6, translated 'heath.' The Arabs call it عَرْعَر. In Persia it is called أَبْهَل.

PINACEAE

J. virginiana *L.*, Red Cedar. Native of Atlantic North America. This species is allied to *J. sabina*, but it is a tree, and its fruits are borne erect on stalks of their own length. So-called cedar-pencils are made of the wood of this tree.

J. excelsa *Bieb.* Native of Greece, Western Asia, and the Himalayas. It grows on Mount Lebanon and has been supposed to be the תִּרְזָה of Is. xliv, 14, 15, which is translated 'cypress.' Our specimen, which is $27\frac{1}{2}$ feet in height, stands in awkward proximity to the great Cedar of Lebanon in the north-east corner of the Garden. It would have been a fine tree if it had had room to grow.

J. drupacea *Labill.* Native of South-east Europe and West Asia. Of all the junipers this species has much the broadest leaves and the largest fruits. These fruits are edible. Their seeds are united together to form a stone, so that the fruit resembles a drupe.

Family GNETACEAE

Of living Gymnosperms the genera *Ephedra*, *Gnetum*, and *Welwitschia* (*Tumboa*), which constitute this family, most nearly approach Angiosperms. Their secondary wood contains vessels, and their flowers have a rudimentary perianth.

Ephedra

Ephedra has about 30 species, native of the warm temperate zones. They all have the xerophilous vegetative form characterising the so-called 'switch-plants' in which the slender green stems play the part of foliage-leaves and the leaves are reduced to minute scales. *Ephedra* externally resembles *Equisetum* and *Casuarina*, but the leaf-scales of *Ephedra* are opposite, and are not arranged in whorls[1].

E. nebrodensis *Tineo.* Native of the Mediterranean Region and North Asia. The great shrub belonging to this

---

[1] *E. altissima* Desf., which climbs up a pillar in the Temperate House, has long, narrow leaves sometimes arranged in whorls of three or four.

THE NEOSIA (*PINUS GERARDIANA*), NEAR THE
ROCK GARDEN (p. 6)

PLATE IV

THE ONE-NEEDLED PINE (*PINUS MONOPHYLLA*),
IN THE PINETUM (p. 6)

GNETACEAE

species, which stands on the west side of the Middle Walk near the Bateman Street end of that walk, was described by Lynch (*loc. cit.*) as the finest Ephedra in the British Isles. It then (1915) measured 21 feet through and 7 feet 8 inches in height. It is now 24 feet 6 inches through, and 7 feet 11 inches high.

E. distachya *L.* Native of Europe and North Asia. This plant is said by some to be the 'homa' of the ancient Zoroastrians, and the 'soma' of the Vedas, about whose identity much has been written. A small specimen will be found beside the great shrub of *E. nebrodensis*, and a larger one on the xerophyte bed near the Lynch Walk. Other species of *Ephedra* grow to the west and south of the Rock Garden.

WELWITSCHIA

Welwitschia has 1 species, of which there is a specimen in the Orchid House, grown from seed collected by the late Professor Pearson.

W. mirabilis *Hk. f.* (*Tumboa bainesii* Wellw.), grows in the arid, stony deserts of Damaraland and Benguella. The structure of this plant is unique. The tap root expands upwards into a short, stout, woody stem, which is surmounted by two opposite leaves. Though the plant may live for longer than a century, yet this pair of leaves, which immediately succeeds the cotyledons, is all the foliage it ever bears. They continue to grow from the base, and to become torn and tattered at the apex. The flowers, which arise on the top of the stem, are most probably pollinated by the insect *Odontopus sexpunctatus*. There is a large dried specimen of this plant in the Museum at the Botany School.

GNETUM

Gnetum has 15 species, all natives of the tropics. They have broad, opposite, coriaceous leaves.

G. gnemon *L.* (in the Stove), a native of the Malay Region, is sometimes cultivated for its edible fruit.

ANGIOSPERMAE

MONOCOTYLEDONEAE

Family GRAMINEAE

ANDROPOGON

Andropogon has over 200 species, cosmopolitan in distribution, but commoner in the warmer parts of the world.

A. sorghum *Brot.* (*Sorghum vulgare* Pers.), Great Millet, Guinea corn, Sorghum. This plant has been grown as a crop from very ancient times. It is supposed to be a cultivated form of *A. halepensis* Sibth. et Smith, a species found growing wild in all warm countries.

Sorghum is a very important crop in many parts of the world. In the Mediterranean Region it replaces the barley and oats of the north. In the drier parts of British India there are on an average 22 million acres of land under this crop yielding annually about 5 million tons of grain.

There are numerous varieties in cultivation. From one of these, chiefly in North America, sugar is obtained; from the panicles of another in the south of Europe they make brooms.

In the Near East the Great Millet is called ذُرة, and in India جوار.

A. nardus *L.*, Citronella grass (in Palm House). This species is cultivated in India. The Citronella oil of commerce, of which large quantities are used in perfumery, is distilled from the leaves.

PENNISETUM

Pennisetum has about 40 species, native of warm regions.

P. americanum *K. Schum.* (*P. typhoideum* Rich.), Bulrush Millet. An important crop in the warmer countries of the Old World. In India, where it is called باجرا, it covers about 14 million acres of land.

GRAMINEAE

Oryza

Oryza has 6 species, native of the tropics. The rice plant will usually be found in the Tropical Aquarium.

O. sativa *L.*, Rice (Tropical Aquarium). Probably indigenous in tropical Africa and Asia. In India, where there are about 54 million acres of land under rice, it has been cultivated from 2800 B.C. Probably more human beings live on rice than on any other grain. It played the principal part in the ancient ceremony instituted by the Chinese Emperor Chin-Nong. In this ceremony the Emperor himself sowed the rice. The other four grains (Wheat, Sorghum, *Setaria italica* Beauv., and Soya bean) were sown by the Imperial Princes.

In India Rice in the field is called دهان, and on the table چاول; the latter is a plural word. In the Repentance of Nasúh we read چاولوں کو ترس گئے, 'we have longed for rice,' and again تاکید کرنا کہ چاول کہڑے نہ رہیں, 'insist upon the rice being thoroughly done.' In the same way 'porridge' in Scotland is sometimes used as a plural word.

Family CYPERACEAE

Cyperus

Cyperus has about 500 species. Most of them are natives of warm countries, but two species grow in the British Isles. The Paper Reed will be found in the Tropical Aquarium, and the Rush-nut in the Corridor.

C. longus *L.*, a British species sometimes called Galingale, grows in abundance on the east bank of the Pond, among *Scirpus maritimus* var. *umbellatus*.

C. papyrus *L.*, Paper Reed. Native of tropical Africa, Calabria, and Sicily. This is the plant of which the ancients made Papyrus. It is now extinct in Lower Egypt, but still grows in the Upper Nile Region, and in Abyssinia. To prepare Papyrus the stems were cut into strips, which, while still moist, were beaten into union with a wooden mallet. The best Papyrus was made of the strips from near

CYPERACEAE

the centre of the stem, and was called Hieratica because it was reserved for religious writings (see Pliny, *N.H.* XIII, 21-23). Paper was first made on a large scale in the eighth century, and by the tenth century it had completely supplanted Papyrus.

The ancients used the unsliced stems of this plant for many purposes. Even small boats were made of them (see Theophrastus, IV, viii, 3, 4). It was probably in such a boat that the infant Moses was exposed among the flags by the Nile's brink.

Papyrus is the Latin name (Greek πάπυρος). Herodotus calls Papyrus βύβλος. In Arabic it is called بردى.

C. esculentus *L.*, Rush-nut, Earth-almond. Native of the Southern Mediterranean Region, and tropical Africa. It has been recorded from India, but was perhaps confounded with an allied species, *C. rotundus* L. (the well-known मोथा), a common weed in all tropical countries. These two, and also other species of *Cyperus*, have fragrant, edible roots.

Family PALMAE

PHOENIX

Phoenix has about 12 species, natives of Africa and the warmer parts of Asia.

Our specimen of the Date Palm is in the Temperate House and *P. sylvestris* is in the Palm House.

P. dactylifera *L.*, Date Palm. This tree has been cultivated for such a long period that it is difficult to tell where it is truly native. De Candolle says that it has existed from time immemorial in the warm, dry zone which extends from Senegal to the Ganges basin, principally between 15° and 20° N. The Date Palm is dioecious, and for thousands of years the people in the East have pollinated the carpellary inflorescence by dividing the staminate spadix and hanging portions of it above the carpellary spathe or actually placing them inside the spathe. One staminate inflorescence is sufficient for from 50 to 100 carpellary trees.

PALMAE

The ovary is apocarpous. Two of the three ripening ovaries fall off early, leaving one to develop into a date.

In Hindustani the Date Palm is کھجور, in Persian درخت‌خرما and in Arabic نَخْل.

اَكْرِمُوا عَمَّتَكُمِ النَّخْلَةَ.

"*Pay respect to your aunt the date palm.*"

P. sylvestris *Roxb.* The commonest palm in India, where it grows in the humid regions in which the Date Palm cannot live. The stems yield sugary juice from which an alcoholic beverage is prepared. The fruits are scarcely edible. Some suppose the Date Palm to be a cultivated form of this tree.

CHAMAEROPS

Chamaerops has 2 species, both natives of the Mediterranean Region. Our specimens of the Dwarf Palm will be found in the Temperate House.

C. humilis *L.*, Dwarf Palm. Native of the Western Mediterranean Region. This is the only palm found wild in Europe, where it is now confined to Spain, though very common on the North African coast. In Europe it is one of the relics of the European Tertiary flora. The ovary, like that of *Phoenix* (Date Palm) is apocarpous, and each berry is derived from a single carpel. Usually only one berry develops in each flower. The leaf-fibres furnish the so-called 'vegetable horse-hair.' The Dwarf Palm multiplies freely by suckers, and in parts of North Africa grows with the luxuriance of a weed. The leaf-bud or 'cabbage' is sold as a vegetable in Marocco and elsewhere.

TRACHYCARPUS

Trachycarpus has about 5 species, natives of East Asia. Our specimen of *T. excelsa* will be found at the entrance of Bay No. 4.

T. excelsa *Wenland*, The Tsung Palm. Native of China. It is cultivated, but probably not indigenous, in Japan. This is the only palm that is really hardy in Great Britain.

PALMAE

It was introduced into Europe in 1830 and into this country in 1836. As no one supposed that it would be hardy, it was at first grown in the tropical palmhouse at Kew, but later trial proved that it needed no protection. It will stand 32° F. of frost. In China the leaf-fibre is used for making hats and the garment called *so-i*, which is worn in wet weather. The Chinese liken this fibre to hair in a horse's mane.

Cocos

Cocos has 36 species, all confined to South America except the Coconut, whose origin is unknown, which is cultivated in all tropical coastal regions. In the Palm House there is a young plant which Mr Preston raised from a Coconut bought in a shop.

C. nucifera *L.*, Coconut Palm. The fruit, which is a great drupe, is derived from all three carpels. The hard 'nut' is the stone of this drupe, which is enclosed in a fibrous layer from which is prepared 'coir' or coconut fibre used for making mats, ropes, etc. The 'kernel' of the nut is the endosperm of the seed, the inner portion of which is liquid (coconut milk). The inner solid part of the endosperm sliced and dried is the 'copra' of commerce, which yields 50 per cent. of fixed oil.

In India the ripe Coconut is called ناریَل. This word and other similar words are all ultimately derived from the Sanskrit नारिकेल. The unripe Coconut is called ڈاب. Its milk, ڈاب کا پانی, is a safe and wholesome drink for thirsty travellers.

The English name is from the Spanish and Portuguese name *Coco*, which means also a monkey's head or bugbear to frighten children. The three pores at the base of the nut suggest this appearance.

Family ARACEAE

The flowers in this family are arranged in a stout spike called a *spadix*, which is more or less ensheathed by a leaf-like bract called a *spathe* (see Pl. V).

ARACEAE

Acorus

Acorus has 2 species, native of the North Temperate Region and South-east Asia. The Sweet Flag grows by the Pond, and on the Medicinal Bed.

A. calamus *L.*, Sweet Flag. Native of Europe, North Asia, the Himalayas, and North America. All parts of this plant, especially the rhizome, are deliciously aromatic. The long, narrow spathe looks like a continuation of the scape, and the spadix appears to be lateral.

The Sweet Flag has been used medicinally from time immemorial. In the old Sanskrit works it is called वच (?'talking'), षड्ग्रन्थ ('six-knotted'), उग्रगन्ध ('strong-smelling') and जटिल ('having tangled hair'). It is the ἄκορος of the ancients, but probably not their 'calamus aromaticus,' which was perhaps *Swertia chirata* (Fam. *Gentianaceae*). In India, where it is still much used, they call it ج‍ـچ, and in Islamic medicine ج‍ـو. Both these names are probably derived from the Sanskrit वच.

The rhizome, though not in the British Pharmacopoeia, is still used medicinally in this country. It contains 1–3·5 per cent. of a bitter, aromatic, volatile oil, and also starch, tannin, and a bitter amorphous principle called Acorin.

Monstera

Monstera has 15 species, all natives of tropical America. A large specimen of *M. deliciosa* will be found on the left as you enter the door of the Palm House.

M. deliciosa *Liebm.* (Pl. V). A root-climber, native of the Mexican rain-forest. At first it is rooted in the ground, but eventually the lower part of the stem dies and the plant becomes an epiphyte with distinct climbing and nutritive roots; the latter, growing down through the air, reach the soil. The leaves when young are entire, but the tissue between the veins grows so slowly that large holes soon appear. Often the tissue between these holes and the edge of the leaf

PLATE V

MONSTERA DELICIOSA, IN THE PALM HOUSE (p. 20)

PLATE VI

THE GOLDEN-CLUB (*ORONTIUM AQUATICUM*), IN THE WATER GARDEN (p. 21)

ARACEAE

breaks away and the leaves become pinnately divided. The ripe infructescences are edible, and in Mexico are sold in the markets. If this strange fruit is eaten before it is quite ripe, the microscopic, needle-shaped crystals of calcium oxalate called *raphides*, which are contained in the cells, pierce the mucous membrane of the mouth and throat, causing great distress. It is during the very short interval between being under-ripe and over-ripe that the fruit justifies the trivial name *deliciosa*. It has a powerful aroma.

Symplocarpus

Symplocarpus has 1 species, the 'Skunk Cabbage,' which grows in the Water Garden.

S. foetidus *Nutt.*, The Skunk Cabbage. Native of Northeast Asia and North-west America. This plant has a thick root-stock whose branches during one year bear large foliage-leaves, and the next year small scale-leaves and flowers. The inflorescences appear in the winter. The thick shell-like spathe is hooded over the globose stalked spadix. The intolerable alliaceous odour of this plant is supposed to resemble that of the skunk.

Orontium

Orontium has 1 species, a specimen of which grows in the Water Garden.

O. aquaticum *L.*, Golden-club (Pl. VI). Native of Atlantic North America. The spathe is very poorly developed; it encloses the spadix when very young but soon tears and remains as a small sheath at the base of the spadix, or falls away altogether. The flowers are hermaphrodite and have a perianth of 4–6 tepals. The starch in the rootstock and seeds is edible after the acrid properties which are probably due to raphides, as in *Monstera*, have been destroyed by heat.

Calla

Calla has 1 species, specimens of which will be found in the Water Garden.

C. palustris *L.*, Water Arum. This plant grows in swampy places in Europe, Siberia, East Asia, and Atlantic North America, but is not found wild in Great Britain. Note the white, open, persistent spathe. Most of the flowers have both stamens and carpels: the very uppermost are sometimes staminate. There is no perianth. Edible starch may be prepared from the acrid rootstock by drying and grinding it, and then heating the powder till it is no longer acrid.

PHILODENDRON

Philodendron has about 200 species, native of tropical America. Our specimens are in the Palm House.

P. erubescens *C. Koch*. The beautiful pink blush on the foliage justifies the trivial name. The arrangement of the veins of the leaves is of more classificatory importance in the *Araceae* than the habit of growth. Note that this plant has the same habit and the same two kinds of roots as *Monstera*, to which however it is not closely allied. In *Monstera* the branches of the midrib give off other veins which break up into a network. In *Philodendron* they all run parallel one with another to the edge of the leaf, giving off no branches, an arrangement characteristic of the sub-family *Philodendroideae*. The well-known 'arum lily' belongs to this sub-family. Its proper name is *Zantedeschia aethiopica* (it is often called *Calla* and *Richardia*). Other well-known members are:—

Dieffenbachia seguine Schott, The Dumb Cane of the West Indies (in the Stove), so called because when pieces of the stems are chewed the resulting swelling of the tongue takes away the power of speech. The plant was formerly used for torturing slaves in this way.

Peltandra virginica Kunth, The Arrow Arum, widely distributed in North America, which will be found in the Water Garden.

COLOCASIA

Colocasia has 6 species, native of the Indo-Malay Region. Our specimens are in the Tropical Aquarium.

ARACEAE

C. antiquorum *Schott*. Wild and cultivated all over tropical India and Ceylon and cultivated in all tropical countries for its tubers, which are good to eat after the raphides in which they abound have been destroyed by boiling. In India the plant is called كچو.

Note the arrangement of the veins of the leaf. The branches of the midrib give off branchlets which unite to form 'collective veins' running parallel with the branches of the midrib. This arrangement is peculiar to the sub-family *Colocasioideae*. Specimens of the genera *Steudnera*, *Alocasia*, *Caladium*, and *Xanthosoma*, which belong to this sub-family, will be found in the Tropical Aquarium, Palm House, and Stove. The tubers of many of them are edible.

PISTIA

Specimens of the only species of this genus will be found floating on the tank in the Tropical Aquarium.

P. stratiotes *L.*, Water Lettuce, is a very common water weed in the warmer regions of both hemispheres. It multiplies freely by runners, often covering the surface of large expanses of water, and sometimes even hindering navigation. As its roots seldom reach the bottom, the plants float freely and are carried about by the wind. The structure of the inflorescence and flowers is so simple that *Pistia* is thought to form a link between the *Araceae* and the *Lemnaceae* (Duckweeds). The larva of a mosquito (*Mansonia titillans*) lives anchored to the roots and obtains its air by piercing them with its tail. The leaves 'sleep' at night. In India the plant is used medicinally.

Family BROMELIACEAE

This family numbers a thousand species, all natives of tropical America. They are particularly abundant in South America. Most of the species do not grow on the ground, but live on the branches of trees, and in the clefts of rocks and cliffs. The leaves of these *Bromeliaceae* are arranged in rosettes. They are narrow, stiff, and thick, and often tipped

with red, and their sheathing bases clasp each other firmly
so as to form a tank in which water and fallen pieces of
plants collect. The roots fasten the plant securely to the
tree or rock on which it grows, but are incapable of
absorbing water. The plant obtains all its water and mineral
food by means of small shield-like scales, which cover the
bases of the leaves and are thus submerged in the tank.
These tanks contain more than water and plant remains;
their flora and fauna are considerable. In some of them even
species of *Utricularia* grow. The terminal inflorescence, which
arises from the centre of the tank, is often very beautiful.

A collection of Bromeliads will be found on either side of
the entrance of the Palm House.

TILLANDSIA

Tillandsia has about 250 species. Many of them are
ordinary tank-plants, but some, particularly the 'Spanish
Moss,' of which there is a specimen at the far end of the
Orchid House, wear a very different aspect.

T. usneoides *L.*, Spanish Moss, Old Man's Beard. This
species extends from the Argentine to Carolina, growing on
trees in huge lichen-like masses. The adult plants have no
roots, and attach themselves to the trees by winding round
the branches, from which they hang down in festoons often
many yards long. Both the stems and the very narrow leaves
are covered with scales like those on the base of the leaves
of the tank-bromeliads. By these scales the plant absorbs
the moisture that condenses upon it. The Spanish Moss is
abundant in its native haunts, where forests draped in it
look very weird. It seldom flowers, but since the wind and
the birds that make their nests of it, carry pieces about from
tree to tree, and these pieces soon grow into huge festoons,
it multiplies rapidly. Spanish Moss is used for packing,
and for stuffing cushions.

PUYA

Puya has about 5 species, native of Peru and Chile.
Unlike ordinary Bromeliads they grow on the ground, and

BROMELIACEAE

have thick, comparatively tall stems. They yield a useful gum, and their stems are used as cork.

P. chilensis *Molina*, a specimen of which grows in Bay No. 7, is native of Chile. The thorn-flanked leaves are arranged in dense crowns at the ends of the branches. In some species the leaves are at the foot of the stem. The huge inflorescence is remarkably beautiful. Our specimen, which is protected during winter, flowered in 1913 and 1920.

ANANAS

Ananas has about 5 species.

A. sativus *Schult.*, the Pineapple, is cultivated in many warm countries. It grows on the ground. The inflorescence is terminal, and the stem grows a short distance beyond it, producing a crown of leaves. The ripe 'Pineapple' is made up of the stem, the bracts, and the fruits, which are berries, consolidated into a fleshy mass. An interesting historical account of this plant will be found in the *Treasury of Botany*, I, 59, Article 'Ananassa.'

Family LILIACEAE

KNIPHOFIA (RED-HOT POKER, TORCH LILY)

Kniphofia has 67 species, native of tropical and subtropical Africa, and of Madagascar.

The Red-hot Pokers are allied to the Aloes, from which they differ chiefly by their terminal (not axillary) inflorescences. Our collection is on the west side of the Broad Walk. The great mass of *K. caulescens* (Pl. VII) is at the entrance of Bay No. 4.

K. uvaria *Hook*. Native of South Africa. This is the common Red-hot Poker of gardens, which has been grown in this country from 1707. Its leaves are two to three feet long, and less than an inch broad.

K. caulescens *Baker* (Pl. VII). Native of South Africa. Most of the species of *Kniphofia* bear their leaves in rosettes on the ground, but this and a few other species have rambling woody stems which curve upwards to bear aloft the crown of foliage and spikes of glowing flowers.

LILIACEAE

Aloë (Aloe)

Aloë has more than 170 species, most of which are native of the steppe-like South African Desert called the Great Karoo. The fleshy leaves are arranged in dense rosettes or crowns, they contain much water, and are entirely devoid of fibres (cf. *Yucca* and *Agave*). The flowers are in axillary racemes. The tubular corolla is often beautifully coloured.

The latest British Pharmacopoeia defines the drug 'Aloe' (Aloes) as the juice that flows from the transversely cut leaves of *Aloë chinensis* Baker, *A. perryi* Baker, and probably other species of Aloë, evaporated to dryness. Former Pharmacopoeias defined Barbados Aloes and Socotrine Aloes as distinct drugs. The Aloes of the Bible (אֲהָלִים, and אֲהָלוֹת) have certainly no connection with the drug Aloes. The Hebrew word probably means the wood of *Aquillaria agallocha* Roxb. (Fam. *Thymelaeaceae*), an Indian tree, whose fragrant wood is burned as incense in the East. The Islamic name for the drug Aloes is صبر, but صَبْر means 'patience.'

صبر تلخ است وليكن بر شيرين دارد

"*Patience (or Aloes) is bitter but has sweet fruit.*"

A. vera L. (*A. barbadensis* Miller), Common Aloe, Barbados Aloe. This is the most northern and the most widely distributed Aloe. It is truly native probably only in North Africa, but has grown from time immemorial in parts of the Mediterranean Region, including Arabia and Syria, and was cultivated in English gardens in Barbados as early as 1596. Its evaporated juice was the 'Aloe Barbadensis' of former Pharmacopoeias. Though probably not native in India this plant must have been known in that country very long ago, as none of its numerous Sanskrit names in any way indicate a foreign origin. Sanskrit writers do not mention the drug Aloes, but preparations of this species are nowadays well-known medicines in India. The Hindi name एलवा (ايلوا) is ultimately from the Greek ἀλόη. Muslims

LILIACEAE

call the drug صَبِر. The Egyptians plant this Aloe in cemeteries, and hang it up over the house-doors to prevent evil spirits entering.

A. perryi *Baker*. Native of Socotra. The dried juice of this species, called Socotrine Aloes, has long been valued as the best kind of Aloes obtainable. The Greeks used it as early as the fourth century B.C., and knew that it came from the Island of Socotra. It is to Bailey Balfour that we owe the discovery that Socotrine Aloes is obtained from this species.

The natives who collect the Aloes dig a small hole in the ground into which they press the middle part of a piece of goat-skin. They cut leaves from the plant and lay them in a circle on the skin so that their severed ends project over the indented part of the skin. The watery sap collected in this way is exported to Muscat in Arabia. It takes about six weeks to harden.

A. plicatilis *Miller*, The Fan Aloe. Native of South Africa. The trunk of this species branches freely, and the strap-like leaves are packed closely together in two rows at the ends of the branches. Our specimen, which occupies the centre of the First Succulent House, must be an old plant. Pl. VIII is from a photograph taken on its journey from the old houses to its present home. It was then a larger plant than it is now. The Fan Aloe was introduced into cultivation at the beginning of the eighteenth century.

A. ciliaris *Haw*. Native of South Africa. This species, of which there is a specimen on your left as you enter the First Succulent House, has long, scrambling stems, which grow very quickly, and bear scattered leaves.

Yucca (Adam's Needle)

Yucca has about 30 species, native of North America and the West Indies. They abound in the South-west United States and Mexico. The woody stems are surmounted by crowns of rigid sword-shaped leaves which differ from those of *Aloë* in not being succulent, and in containing fibre. The absence of conspicuous teeth in the margins distinguishes them from those of most species of *Agave*. The large, white, globular drooping flowers resemble those of neither genus. Our collection is on the west side of the Broad Walk.

Y. gloriosa *L.* Native of eastern North America, where it grows near the coast and often on sand dunes. This is the best-known species. Its bulky stem rarely branches. The leaves are up to three inches wide, and quite straight.

Y. recurvifolia *Salisb.* Native of the south-eastern United States, where it grows near the coast. It is distinguished from *Y. gloriosa* by its freely-branched stem and recurved leaves.

Y. glauca *Nutt.* (*Y. angustifolia* Pursh.). Native of the South and Central United States. The stem of this species does not rise above the ground, and the very narrow leaves are arranged in a graceful spherical head.

Y. australis *Trel.* (*Y. filifera* Chab.). Native of Mexico. The specimen of this plant in the Temperate House has a stem $12\frac{1}{2}$ feet high. The margins of the leaves of this and several other species of Yucca give off fine white threads. In its native land the inflorescences of this species are said to look like gleaming waterfalls pouring out from the ends of the branches (see *Botanical Magazine*, T. 7179).

Y. elephantipes *Regel.* (*Y. guatemalensis* Baker). Native of Guatemala and Mexico. The specimen of this species in the Palm House, to judge by the thickness of its trunk, must be very old.

Family AMARYLLIDACEAE

Agave (Century Plant, American Aloe)

Agave has 50 species, native of tropical America and the southern United States. Several yield good leaf-fibre. During the sixteenth, seventeenth and eighteenth centuries the useful and ornamental species were carried to all the warmer regions of the world. The plants live for many years before they flower and die. The copiously-branched inflorescence of some species attains 20 feet in height. The

PLATE VII

A MASS OF RED-HOT POKER (*KNIPHOFIA CAULESCENS*), AT THE ENTRANCE OF BAY No. 4 (p. 25)

PLATE VIII

THE GREAT FAN ALOE (*ALOE PLICATILIS*) ON ITS WAY TO ITS PRESENT HOME (p. 27)

AMARYLLIDACEAE

Mexicans prepare their national drink, *pulque*, by cutting the young scape of *A. salmiana* Otto, and collecting the sap that flows from it. A vigorous plant will yield 4 to 5 litres daily, sometimes giving a total yield of 1100 litres. The species of *Agave* naturalised in the Mediterranean Region is *A. vera-cruz* Miller. Our collection is in the Succulent House. Specimens of *A. parryi* Englm. and *A. utahensis* Englm. grow between Bays 1 and 2, where they are protected in winter.

BOMAREA

Bomarea has 50 species, native of South America. They are twining plants. The leaves are usually borne on short, twisted stalks, so that the original lower surface faces upwards. The flowers are arranged in pendulous umbel-like inflorescences. We have a good collection of Bomareas, of which the following are of interest.

B. carderi *Mast.* Native of New Granada. A specimen of this species grows in the Corridor. Its magnificent flowers are about two inches long, rose-pink without, and dark purple spotted within. They are arranged in long loose inflorescences.

B. patacocensis *Herb.* Native of the Andes of Ecuador and of Columbia. It was discovered at a place called Patacocha. This species luxuriates in the First Succulent House, where it climbs aloft and flowers very freely. The flowers are bright red and arranged in profuse inflorescences which look like hanging Japanese lanterns. The outer perianth segments are much shorter than the inner ones.

B. cantabrigiensis *Lynch* (Pl. IX). This is a hybrid between *B. caldasiana* Herb. and *B. hirtella*. (Both parents grow in the Corridor.) It grows in Bay No. 4, and should be quite hardy in the west of England.

Family MUSACEAE

Nearly all the plants of this tropical family are tree-herbs (see *Musa*). Their large oval leaf-blades tear very readily from the margin right up to the stout midrib.

MUSACEAE

Musa (Banana, Plantain)

Musa has about 30 species, native of the tropics of the Old World. They are all tree-herbs. The creeping rootstocks give rise to groups of leaves whose basal sheathing portions are so wrapped together as to produce the appearance of a stout cylindrical stem. The large inflorescence springs directly from the rootstock and appears at the summit of this apparent stem. Each bract contains in its axil a row of flowers. The rows of carpellary flowers are at the base of the inflorescence and give rise to the 'hands' of bananas sold in the shops. There are sometimes some hermaphrodite flowers between the carpellary ones and the staminate ones, which are toward the end of the inflorescence. Most of our specimens are in the Palm House and Tropical Aquarium.

M. sapientum *L.*, The Banana. Native of the eastern hilly districts of India, and of Burma. It is cultivated throughout India and the tropics. The banana gives far more food per acre than any other crop. The core of the fruit when dried yields 40 per cent. of meal which may be kept for any length of time. Indians use the fibre of the leaf-base for making cord, mats, and coarse paper, but it is not nearly such a strong fibre as Manila Hemp, which is obtained from *M. textilis* Née, a species much grown in the Philippines.

M. ensete *Gmel.*, Abyssinian Banana. Native of East Africa. In its native land this species grows to a height of 40 feet. The leaf-blades, which are traversed by bright-red midribs, measure nearly 20 feet in length and 3 feet across. The fruit is dry and inedible, but the stalk of the inflorescence is cooked and eaten.

It has often been stated that this plant is figured on ancient Egyptian monuments, but Professor Percy Newberry tells me that he has neither seen any representation of it, nor heard of any records of its fruit or leaves being found in Egyptian tombs, and that he does not believe that

MUSACEAE

the plant was known to the ancient Egyptians. The myth probably has no surer foundation than the tales about the germination of the so-called 'mummy wheat.'

M. cavendishii *Lamb.*, Chinese Dwarf Banana. Native of South China. This species is extensively cultivated in the West Indies and along the coast of the southern United States. Large quantities of its fruits are sent from Jamaica and other West India Islands to Europe.

M. basjoo *Sieb. et Zucc.*, Japanese Banana. Native of the Liu Kiu Archipelago (between Japan and Formosa). In the south of Japan it is cultivated for its leaf-fibre. This is probably the hardiest species of *Musa*. With us it grows at the entrance of Bay No. 5, but it is protected with rushes during winter.

Family ZINGIBERACEAE

ELETTARIA

Elettaria has 2 species, native of India. Our specimens are in the Stove.

E. cardamomum *Maton*, The Cardamom Plant. Native in the moist forests of the hilly districts of West and South India. It is cultivated in these and other districts. The deliciously aromatic seeds are the Cardamomi semina of the British Pharmacopoeia. Large quantities are annually imported into Europe, where they are used for flavouring and medicine. They are a favourite condiment in the East. The little three-cornered capsules that contain the seeds are never plucked, as the pressure of the fingers is liable to burst them. They are collected by cutting the stalks by which they are attached to the inflorescence.

In India Cardamoms are called الائچی (the Sanskrit name is एला), and in Persia هل.

DICOTYLEDONEAE

ARCHICHLAMYDEAE

Family CASUARINACEAE

The 25 species of *Casuarina*, the only genus of this family, are natives of Australia, Polynesia, and the Indo-Malay Region. (*C. equisetifolia* Forst. is planted in many tropical countries.) They are all switch-plants with deeply-furrowed stems and whorls of scale-leaves (see *Ephedra*, p. 12). Some botanists suppose that they are the most primitive living Angiosperms. They resemble Gymnosperms in several anatomical details. Some specimens of *Casuarina* will be found in the Temperate House.

Family SALICACEAE

Populus (Poplar)

Populus has about 30 species, native of the North Temperate Zone. Our collection of Poplars will be found on the south of the Border Walk.

P. alba *L.*, White Poplar, Abele. Native probably from South-east Europe to Turkestan. It is much planted in Europe and Britain. It is the λεύκη of Theophrastus and Dioscorides, the *leuce* of Virgil, Horace, and Pliny, and the سفیدار (white tree) of the East. These names refer to the white bark and snow-white felting of hairs on the lower surface of the palmately-lobed, somewhat maple-like leaves. This 'tomentum' checks excessive loss of water and is often present on the leaves of plants which live in dry regions. Of our Poplars of the section *Leuce* this species is native of the driest region, and has the densest covering of hairs. The tomentum of the Grey Poplar (*P. canescens* Smith), whose area of distribution extends to Western Europe, is thinner and tends to disappear as the leaf grows older. We have in Britain two varieties of the Aspen (*P.

SALICACEAE

tremula L.). In var. *sericea* the leaves, when unfolding, are covered with long silky hairs. This is the common variety in the drier parts of Great Britain and is often seen in Cambridgeshire. Var. *glabra*, whose leaves are almost without any trace of hairs, is the commoner, if not indeed the only form of the Aspen, in the hilly and rainy districts of western and northern Great Britain and Ireland.

P. canescens *Smith*, Grey Poplar. Native of Europe and Asia Minor. It is indigenous, but not common, in southern and central England. This is perhaps the handsomest member of the British flora. The bark of the trunk is brown and furrowed; that of the boughs, which are very massive, greyish white and smooth. The leaves are less maple-like than those of *P. alba*, and approach more nearly the circular outline of those of *P. tremula*, the Aspen, but they are more coarsely toothed than Aspen leaves, and except when old, are covered with grey felt beneath. The Grey Poplar affects the upland borders of fens.

P. tremula *L.*, Aspen (earlier *asp* was the noun, and *aspen* the adjective). Native of Europe, North Africa, and West Asia. It is found throughout the British Isles. The leaf-blades are nearly circular, usually without hairs (see above) and set on long, flattened stalks. Aspen leaves tremble more actively and more audibly than those of other poplars.

P. italica *Moench*, Lombardy Poplar. Native land not known. It is certainly not indigenous in Lombardy. It is perhaps the best-known Poplar in this country. Fastigiate forms of trees are usually varieties of species whose branches spread in an ordinary manner, and it is often supposed that the Lombardy Poplar is a fastigiate variety of the Black Poplar (*P. nigra*). Moss (*C.B.F.* xi, 9) strongly opposes this view. "The statement is commonly made that the Lombardy poplar differs from the black poplar only in habit; but we find the differences of the two plants to be indefinite in number. These differences apply to the habit, to the shape of the buds and leaves, to the time of unfolding of the leaves and of the catkins, to the time of leaf-fall, to

SALICACEAE

the time of flowering, and to the structure of the different parts of the flower. We have no hesitation therefore in regarding *P. italica* and *P. nigra* as distinct species. If *P. italica* is merely a fastigiate form of *P. nigra*, we can only say that it is a fastigiate form of some variety of *P. nigra* which we have never seen."

× **P. serotina** *Hartig*, Black Italian Poplar. This tree, which is supposed to be a hybrid between the American *P. deltoidea* Marshall, and the British *P. nigra* L., is overwhelmingly the commonest Poplar in the British Isles. It is one of the tallest and fastest growing of our trees, and bears a huge, somewhat fan-like crown of curved, ascending branches. Long after the other poplars have expanded their leaves the Black Italian Poplar slowly unfolds its bronze-coloured young foliage (Latin *serotinus*, late). Anyone who cannot distinguish the crown of this tree from that of our other trees has not a full entrance into Cambridgeshire scenery[1]. It is always a staminate tree and is propagated by cuttings. When the breeze in summer shakes its leaves they make a noise like rain falling. Its sharply-pointed ovate leaves are unlike the circular leaves of the Aspen, with which it is often confused.

SALIX (WILLOW)

Salix has 170 species, most of them native of the North Temperate and Arctic Regions. They hybridise very freely. The glucoside *salicin* is obtained commercially from the bark of *S. fragilis*, *S. purpurea* and other species. Our collection, of which the greater part consists of hybrids, is between the Border Walk and the Pond. A new collection (of species) is being started on the west side of the Pond. The rosette-like galls often seen on Willows are sometimes called Butterfly Galls. They are not caused by a butterfly, but by the dipterous insect *Cecidomyia rosaria*. The follow-

---

[1] *P. deltoidea*, with which it might be confused, is now rarely seen in Britain. Typical *P. nigra* is not uncommon in Cambridgeshire. Its huge trunk and massive *spreading and descending* boughs usually bear large corky bosses.

SALICACEAE

ing notes are intended to be an introduction to the commoner Cambridgeshire Willows.

S. alba *L.*, White Willow. Native of Europe, North Africa, and parts of Asia. In Britain it is common in lowland localities by stream sides, in wet alluvial meadows, marshes, and fens. It is very commonly planted for pollarding[1]. When left to itself the White Willow becomes a large tree, sometimes attaining a height of nearly 100 feet, with thick, rugged bark. It is easily recognised by its sharp-pointed leaf-blades five or six times as long as broad, and covered densely beneath and sparsely above with white, silky hairs. The foliage is of striking beauty when moved by the wind on a summer day. The timber of var. *caerulea* is used for cricket bats. Var. *vitellina* has red or orange twigs.

S. fragilis *L.*, Crack Willow, Withy. Native of Europe, and of North and West Asia. This species differs from the White Willow in its wider angle of branching, in its larger, more deeply serrate leaves, which are quite devoid of hairs when mature, and in the shining (not silky) young twigs breaking off easily with a snap. In Britain, as an indigenous tree, it is commoner than the White Willow and thrives in soils poorer in mineral content.

S. babylonica *L.*, Weeping Willow. Native of China. Planted in many lands. It may be known by its long, pendulous young branches. The leaves are like those of the White Willow and the Withy, but are even longer in proportion to their width. Our specimen is said to have been grown from a cutting from the tree growing on Napoleon's grave in St Helena. In var. *annularis* the leaves are rolled backwards into a spiral. In Persia the Weeping Willow is called بید مجنون.

S. caprea *L.*, Goat Sallow, Palm. Native of Europe and Northern Asia, extending to Japan. In Britain it is common in woods and hedgerows. This species is rarely more than a large shrub, and can be distinguished from the three

[1] Willows are pollarded, or cut off at about 8 feet from the ground, in order to produce an abundant growth of young branches, which are used for making hurdles.

preceding species by its leathery leaf-blades being only about twice as long as broad, or sometimes almost circular in outline. Its twigs are smooth, hairless, and often ruddy. This species and *S. cinerea* are the earliest flowering British Willows and their catkin-bearing twigs are the 'Palm branches' of Palm Sunday.

S. cinerea *L.*, Sallow. Native of Europe, North Africa, the Caucasus, and Western Asia to Kamtchatka. This is the common Sallow of the fens, and indeed the commonest and most widely distributed of all the British Willows. It is closely allied to the Goat Sallow and differs from it in its blackish, very hairy twigs, and narrower, more wrinkled leaf-blades.

Other common Cambridge Willows are: *S. viminalis* L., the Common Osier, which has leaf-blades about twenty times as long as broad, with netted upper surface, and shining white lower surface, and wavy, recurved margins[1]. *S. purpurea* L., the Purple Osier, which differs from all our other Willows in often having its narrow, thin, glaucous leaves arranged nearly or quite opposite each other on the twigs, and in its exceedingly bitter bark[2]. The name *Osier* is given to Willows whose tough, pliant branches are used for basket-making. They are often grown in beds and coppiced. *S. repens* L., Creeping Willow, which grows in the fens (the fen variety does not creep), rarely attains more than a yard in height. It has small leaves about an inch in length brightly shining above, and silky beneath. Rather an uncommon Willow about Cambridge is *S. triandra* L., the Almond-leaved, or French Willow. Its bark flakes like that of a Plane. The leaves are variable in shape but are dark green and shining above, and usually bear two glands at the junction of the stalk and blade. The stipules of this species are very large.

[1] Hybrids of *S. viminalis* with *S. caprea* and *S. cinerea*, distinguished from the first parent by their broader much more hairy leaves, are commonly planted as osiers.

[2] Bitter from the abundance of *salicin* it contains.

PLATE IX

BOMAREA CANTABRIGIENSIS,
IN BAY No. 4 (p. 29)

PLATE X

BOLE OF THE PAPER BIRCH (*BETULA PAPYRIFERA*) (p. 41)

MYRICACEAE

Family MYRICACEAE

MYRICA

Myrica has about 50 species, natives of the temperate and warmer parts of both hemispheres. Some species, for example *M. nagi*, have edible fruits. *M. gale* and *Comptonia* are by the stream outside the entrance to the Water Garden. *M. cerifera* is on the east side of the East Walk.

M. gale *L.*, Bog Myrtle or Sweet Gale. Native of various parts of Europe, of North Asia, and of North America. In Great Britain it is locally abundant on wet, siliceous hill slopes and on moors. On fens it is rather rare, but it grows on Wicken Fen. The whole plant is aromatic and the leaves are used to scent clothes and to keep away vermin.

M. cerifera *L.*, Wax Myrtle. Native of eastern North America, Bahamas, Bermudas, etc. The thick layer of pale blue wax which covers the fruit of this and other species is removed by boiling. The early settlers in America made their candles of it.

COMPTONIA

Comptonia has a single species which is often included in *Myrica*.

C. asplenifolia *Gaertn.*, Sweet Fern. Native in the eastern and northern United States, where it grows in dry, sterile soils. The leaves are pinnately divided and look something like fern fronds.

Family JUGLANDACEAE

All our Juglandaceae will be found in the neighbourhood of the north-west corner of the Garden.

PTEROCARYA

Pterocarya has 7 known species. One of these is native of Japan, 5 of China, and 1 of the Caucasus and Persia.

Our specimen of *P. fraxinifolia* forms a great thicket

JUGLANDACEAE

through which is cut the course of the little stream that feeds the Pond (Frontispiece). Originally two trees stood here, but about the year 1886 one blew down, and many of the suckers which are constantly springing from its roots are now large trees.

P. fraxinifolia *K. Koch* (*P. caucasica* C. A. Meyer), Caucasian Wing-nut. Grows in moist places in the Caucasus and in Persia. Notice the chambered pith in the twigs which among *Juglandaceae* is characteristic of the genera *Pterocarya* and *Juglans*. The buds have no protective scales, but are covered with rusty down. The two bracteoles form a wing round the fruit.

JUGLANS

Juglans has 15 species, natives of Europe, Asia, and North and South America.

J. regia *L.*, Walnut. The Walnut has been cultivated for such a long time in so many different countries that it is difficult to tell where it is indigenous. It is generally said to be native from the east of Europe and Asia Minor to Afghanistan; but it is found apparently wild as far east as Assam, and may possibly be native even in China and Japan. The well-known nut is really no nut at all, but the stone of a drupe. The hard shell of this stone splits down the midrib of each of the two carpels into two 'boats.' Walnuts were well known to the Romans, who called them *nux* in their poetry, and *iuglans* in their prose. The word *iuglans* is from *iovis glans* (Jupiter's Acorn). The Greeks used the word Διὸς βάλανος for the Walnut, the Sweet Chestnut, and the Hazel nut.

The Romans used to throw Walnuts at weddings ('sparge marite nuces,' VIRG.). Such passages as 'da nuces pueris iners,' CAT. and 'nucibus relictis,' PERS. ('putting away childish things') teach us that Walnuts were considered good and safe toys for Roman children.

Walnuts in Persian are called چهار مغز ('four kernels') and also گردو and گردگان (from درد, round). The Persian equivalent of our 'All is not gold that glitters' is 'Everything that is round is not a Walnut' (هرچه گرد است گردو نیست).

JUGLANDACEAE

Sa'di compares educating those of bad origin to placing a Walnut on a dome.

پرتو نیکان نگیرد هرکه بنیادش بدست *
تربیت نا اهل را چون گردگان بر گنبدست

The leaves are fragrant when crushed and have about seven, rather broad, almost entire leaflets.

J. nigra, Black Walnut. Native of eastern and central North America. The leaves have twelve or more narrow, serrate leaflets. Our specimen is a noble one. The timber of both these species is valuable and well-known. Compare carefully the bark of these two trees.

Family BETULACEAE

CARPINUS

Carpinus has about 18 species, native of the North Temperate Zone. Our best hornbeam is by the path near the maples.

C. betulus *L.*, Hornbeam. Native of Europe (where it extends to 57° N.), and of West Asia. It is indigenous in the south-east of England on sandy and clayey loams containing a low proportion of lime. As the hornbeam tolerates frequent cutting it is much exploited as coppice. The small nut lies at the base of a persistent, 3-lobed, leaf-like involucre. The timber is remarkably hard, and is used for wooden screws, cogs, and pianoforte keys, and makes excellent fuel. The name Hornbeam refers to the hardness of the wood (O.E. *horn*, horn, and *béam*, tree).

CORYLUS (HAZEL)

Corylus has about 8 species, natives of the North Temperate Zone. Our hazels are on the east side of the East Walk.

C. avellana *L.*, Hazel. Native of Europe, West Asia, and (?) North Africa. It is found throughout the British Isles, and forms the bulk of the shrubby undergrowth in the oak woods and ash-oak woods in southern England, and is by far the commonest component of our coppiced woods. It ascends to nearly 1900 feet in the Highlands. Latin authors write of *corylus*, but make no mention of its nuts. Since both the goat and the hazel were enemies of the vine, they used to roast goat's entrails on hazel spits: 'in uerubus torrebimus exta colurnis,' VIRG.

In Persian the hazel is called درختِ فُندُق and the nuts are called چلغوزه (cf. *Pinus gerardiana*, p. 6).

C. tubulosa *Willd.* (*C. maxima* Mill.), Filbert. Native of the Eastern Mediterranean Region. Introduced into Britain in 1759. This is a larger, more tree-like plant than *C. avellana*. The cup is twice as long as the nut and prolonged beyond it into a tube.

C. colurna *L.*, Turkish Hazel. Native of South-east Europe and of the Himalayas. This species is a large tree. It is locally very abundant in the north-west portions of the Himalayas at altitudes of from 5500 to 10,000 feet. The nuts, which are smaller than our hazel nuts, are eaten all over India. The best kinds come from Afghanistan and Kashmir. They are densely clustered in groups of three or more. The edge of the cup is prolonged into numerous, narrow, pointed segments. In India spinning wheels are made of the wood of this tree.

BETULA (BIRCH)

Betula has 40 species, natives of the North Temperate and Arctic Zones. The collection of Birches extends from the east side of the East Walk into the south-east corner of the Garden.

Three Birches are natives of the British Isles. Specimens of the two commonest species (*B. alba* and *B. tomentosa*) will be found in the general collection. The Dwarf Birch (*B. nana*) thrives on a dry bank in the Water Garden.

B. alba *L.* (*B. verrucosa* Ehrh.), White Birch. Native of Europe (extending to 65° N. in Sweden) and of parts of Asia and North America. The young branches bear resinous, peltate glands, and no hairs. The leaf-blades are triangular or rhomboid, doubly serrate, and acuminate. The lateral lobes of the bracts are more or less falcate. This is the more frequent birch of limestones, and occurs on the southern chalk more commonly than *B. pubescens* Ehrh., but both species (with hybrids) frequently grow together. (Compare *B. pubescens* Ehrh.) The timber of both these birches is among the best European woods for turnery.

B. pubescens *Ehrh.*, Common Birch. Native of Europe, where it extends to 71° N., and of Asia and North America. The young branches have no resinous glands but are hairy. The leaf-blades are ovate, cordate at the base, and singly or irregularly serrate. The lateral lobes of the bracts are spreading or erect. This species is characteristic of upland siliceous soils in the west and north of Britain, where it ascends to 760 m. (2500 feet), sometimes forming a zone of birch woods above the oak woods. It is the commonest tree in Scotland, especially in the north, but is also common in Britain generally. Compare *B. alba* L. Many varieties of *B. pubescens* have been described.

B. nana *L.*, Dwarf Birch. The Dwarf Birch is a small bush whose leaves are almost circular, or sometimes even broader than long, and have crenate margins. It is essentially an arctic plant, and during the Ice Age spread widely over Europe. It still occurs on peat as a glacial relic in the extreme north of England, the eastern half of Scotland, and in many places in Central Europe. Its leaves have been found embedded in peat at Barnwell, near Cambridge, together with those of Arctic Willows.

B. papyrifera *Marsh*, Paper Birch, Canoe Birch (Pl. X). Native of North America where it reaches as far north as Labrador and Hudson's Bay and extends south to Iowa and Nebraska. It occurs scattered through forests of other trees and affects rich wooded slopes and borders of streams, lakes, and swamps. The leaves are much larger than those of our Birches. The Northern Indians use the tough, resinous bark for making canoes, baskets, drinking cups, and other small articles. In winter they often cover their wigwams with it. A canoe made of this bark weighing forty to fifty pounds will carry four persons. The bark, like that of other Birches, peels off in layers and is marked with long horizontal lenticels. The timber of this species is used for spools, shoe-lasts, pegs, and generally in turnery. It makes good wood pulp and excellent fuel. The Paper Birch was introduced into this country in 1750. Our specimen has been grafted on to a stock of another species and the union between the stock and scion is apparent about a foot from the ground.

ALNUS (ALDER)

Alnus has 17 species, native of the North Temperate Zone, the Mediterranean Region, and the Andes. The ripe carpellary catkins of the Alders are hard and woody, and resemble the cones of *Sequoia*.

A. glutinosa *Gaertner*, Common Alder. Native of Europe, from the Caucasus to Japan, and of North Africa. It occurs throughout the British Isles by stream sides, in alluvial meadows, and in fens, never thriving unless its roots are supplied with well-aerated water. The wood, though soft, lasts a long time in water. The Dutch use it for making piles for bridges. In former times it provided good charcoal for making gunpowder. The mature leaves are obovate, or almost circular in outline, and green and hairless on both surfaces.

A. incana *DC.*, Grey Alder. Native of Europe (not of Britain) and of eastern North America. It extends to $70\frac{1}{2}°$ N. in Scandinavia. The mature leaves are much narrower and more acute than those of the Common Alder; the lower surface is bluish-green, and more or less hairy. This species is naturalised in the district of Furness, where it has been planted in wind-screens, and in other parts of Britain.

A. cordifolia *Tenore*, Italian Alder. Native of Corsica and South Italy. This species is a large tree with handsome, heart-shaped leaves, and 'cones' an inch or more in length.

FAGACEAE

Family FAGACEAE

FAGUS (BEECH)

Fagus has about 4 species, native of the North Temperate Zone. There are several specimens of the Beech in the Garden.

F. sylvatica *L.*, Beech. Native of Europe. The Beech, which is a surface-rooting tree and only flourishes on light and well-drained soils, is the dominant tree of the woods (including the so-called 'Hangers') on chalk soil in the south-east of England and the calcareous oolites of Gloucestershire. It also occurs on sandy soils. Towards the north and west it is often planted, but is not indigenous. The green wood is excellent fuel, and is much used in turnery, as in the old-established chair-making industry at High Wycombe in Buckinghamshire. The Beech was well known to the Romans, who describe its dense shade and smooth bark:—

inter densas, umbrosa cacumina, fagos, (VIRG. *Ecl.* II, 3.)

and

incisae seruant a te mea nomina fagi...
et quantum trunci, tantum mea nomina crescunt.
(Ov. *Her.* V, 21.)

CASTANEA (SWEET CHESTNUT)

Castanea has 30 species, natives of the North Temperate Zone, and of tropical India. The nuts are in groups of three enclosed in a prickly covering (cf. *Aesculus*, p. 82).

C. sativa *Miller*, Sweet Chestnut, Spanish Chestnut. It is difficult to tell where this tree is truly native. It has been cultivated in Italy from the time of Virgil. Chestnut woods abound in the Mediterranean Region from Asia Minor westwards, and chestnuts are a very important food in many countries, especially Spain and Portugal. In south-eastern England the tree is common, but doubtfully indigenous. In Kent, where it grows in woods on sandy and gravelly soils, the coppiced branches are used as hop poles. The timber resembles that of *Quercus sessiliflora* Salisb.

FAGACEAE

Quercus (Oak)

Quercus has 200 species, native of the northern hemisphere. Most of our oaks will be found near the north-east corner of the Garden.

Q. robur *L.* (*Q. pedunculata* Ehrh.), Common Oak. This is by far the commonest tree in England, but it does not reach the north of Scotland. It is native of Europe, where it has a very wide range, extending northwards to 62° 55' in Norway, and of Western and South-western Asia. This spécies flourishes best on moist, deep, and rather heavy soils. The mature leaf-blades are glabrous and have reflexed auricles at the base. The petioles are usually short or absent. The peduncles, which bear one or more acorns, are long and slender. Compare *Q. sessiliflora* Salisb.

Q. sessiliflora *Salisb.*, Sessile-fruited Oak or Durmast. Native of Europe, where it has a less extensive range than *Q. robur* L., and eastward to Persia. This species grows especially on siliceous and leached soils. It is the characteristic species of the hill slopes of the west and north of Britain, but does not reach the north of Scotland. The mature leaf-blades are flatter than those of *Q. robur* L., and have shallower lobes. They have bifid or multiple hairs beneath, and no auricles at the base. The petioles are usually much longer than those of *Q. robur* L. The groups of acorns are sessile or nearly so, and the short peduncles, where present, are stout.

Q. cerris *L.*, The Turkey Oak. Native of Southern Europe and Asia Minor. The oaks of the subsection *Aegilops* of the section *Lepidobalanos* have deciduous leaves which are hairy beneath. The fruits usually take two summers to ripen[1]. The scales of the cupule are linear and reflexed. The Turkey Oak, known by its long, narrow, deeply-cut leaves, and thread-like bud-scales, is naturalised on dry sandy soils in southern England, and self-sown trees are locally abundant in Bedfordshire and Cambridgeshire. The timber is of little value.

Q. suber *L.*, The Cork Oak. Native of the Mediterranean

[1] The Turkey Oak in this country commonly ripens its fruit in a single summer.

PLATE XI

THE OAK BESIDE THE BROAD WALK (p. 45)

PLATE XII

ASIMINA TRILOBA, FLOWERING SPRAY (p. 55)

FAGACEAE

Region. The oaks belonging to the subsection *Suber* of the section *Lepidobalanos* have evergreen leaves densely tomentose beneath. The fruits ripen usually in a single summer. The scales of the cupule are appressed or erect. This species is readily distinguished by its thick bark, which is the 'cork' of commerce. The chief supplies of cork come from Portugal. The trunk and chief branches of the trees are stripped every eight or ten years. The leaves have 5–7 pairs of lateral nerves.

Q. ilex *L.*, Evergreen Oak, Holm Oak. Native of the Mediterranean Region. This species differs from *Q. suber* in its bark which produces no cork, in its leaves which have 6–10 pairs of lateral nerves, and in the cup of its acorn, which is more rounded and less conical than that of the Cork Oak. It is often planted in this country, and in the south-west of England is quite naturalised. The leaf is a typical sclerophyll.

The Oak (Pl. XI) that stands on the west side of the Broad Walk, near to its junction with the South Walk, has been identified by Henry and Elwes as *Q. obtusata* Humb. et Bonpl., but it does not agree with the original description and figure of this species in *Plantae aequinoctiales* (II, 26, t. 76). At present the identity of our plant baffles determination. If it be a hybrid, probably one of its parents is the Common Oak (*Q. robur* L.), which it strongly resembles in its acorns, and the stalks on which they are borne. It differs from the Common Oak in its leaves, which are nearly evergreen, and are borne on longer stalks. The true *Q. obtusata* shares with *Q. lanceolata* the peculiarity of having all the scales of the acorn-cup pointing downwards (towards the stalk). Both are natives of Mexico.

Family ULMACEAE

ULMUS (ELM)

Ulmus has about 20 species, native of the North Temperate Region. Our collection is in the south-west corner of the Garden.

ULMACEAE

U. campestris *L.*, English Elm. The native land of this tree is unknown. It is very common in the lowlands of southern England, and in the western Midlands. It is rare in Scotland, where it only attains half its natural height. Moss thinks that it was a constituent of our original forests on damp soils and especially on alluvial deposits. I know of no evidence to support the statement that the Romans introduced it into England. It is said to be indigenous in Holland and Spain. All Englishmen know this tree by sight. The lower branches, which are naturally very large and widely spreading, are almost always lopped. The tall, straight trunk is more or less enwrapped in a maze of short adventitious branches. The boughs end in heavy masses of dark green foliage which show like cumulus against the summer sky. The leaf-stalks are rather long, and bear medium-sized blades which are rough above. The great Elms along the 'Backs' are of this species.

U. nitens *Moench*, The Smooth-leaved Elm. This species is indigenous in eastern England, in many parts of Europe, in Northern Africa, and in parts of Asia. It is perhaps the commonest tree in the hedgerows of Cambridgeshire. You may distinguish it from our other Elms by the mature leaf-blades which are borne on long stalks, and are strikingly unequal at the base, glabrous on both surfaces, and very smooth and shining above. The lower branches spread widely, and the terminal branchlets often droop. The foliage is much paler than that of *U. campestris*.

U. glabra *Hudson* (1762) (*U. montana* Stokes (1787)), Scotch, Wych, or Mountain Elm. Native of Europe and of Northern Asia to the Amur region. It is indigenous throughout Great Britain, but is much commoner in the north and west. This tree can be readily distinguished from our other Elms by the large, very rough leaves whose stalks are so short that they are often hidden by the bases of the blades. The branches are long and spreading and often droop at the extremities (the word 'wych' originally meant 'drooping').

× **U. vegeta** *Schneider*, The Huntingdon Elm. This tree,

ULMACEAE

which is said to have been raised from seed in a nursery at Huntingdon in the middle of the eighteenth century, is probably a hybrid between *U. nitens* Moench (the Smooth-leaved Elm) and *U. glabra* Hudson (the Wych Elm)[1]. The long branches ascend at an acute angle from a rather short bole. The leaves have the shape and size of those of *U. glabra* (but not their short stalks) and the smooth surface of those of *U. nitens*. The trees in Brooklands Avenue are Huntingdon Elms.

U. stricta *Lindley*, The Cornish Elm, whose origin is unknown, is distinguished by its short, ascending branches, and narrow, pyramidal outline. The leaves resemble those of *U. nitens*, but are smaller, and less oblique at the base, where they show a concavity above.

U. sativa *Miller*, The Small-leaved Elm[2], is probably native of Western, Central and Southern Europe, and Western Asia. It occurs locally in southern England, chiefly in the Eastern Counties. Its leaves, which are borne on slender, interlacing branchlets, are much smaller than those of our other Elms.

Family MORACEAE

All the Moraceae have a milky juice. The small and inconspicuous flowers are usually crowded together into very dense inflorescences. These inflorescences, owing to the intercalary growth of the axis, often assume very curious shapes.

Morus (Mulberry)

Morus has 10 species, native of the North Temperate Regions. Each 'mulberry' is derived from a whole spike-like inflorescence. Compare it with a Blackberry (*Rubus*), which is derived from a single flower. Leaves of very

---

[1] The Dutch Elm (*U. hollandica*, *U. major*) is thought to have the same parents. Its wide-spreading lower branches distinguish it from *U. vegeta*.

[2] Usually called *U. minor*. It is generally held that Miller's *U. sativa* was the English Elm.

MORACEAE

different forms are often seen on the same branch of the Mulberries. The stipules are small. Our collection is in the south-west corner of the Garden.

M. alba *L.*, White Mulberry. Native of China and perhaps of the north of India. The 'mulberries' of this species are white or pinkish and insipid, and are usually borne on distinct stalks. The White Mulberry is cultivated in many lands, chiefly for its leaves upon which silkworms are reared; it has been grown in the Mediterranean Region from the twelfth century. Its Chinese name is *sang*, and in Persian it is called توت سفید.

M. nigra *L.*, Black Mulberry. Native of Persia. Now cultivated in many regions. The fruits of this species are almost black when ripe, and of delicious flavour. Their stalks are short or absent. In Persia the Black Mulberry is called شاه توت. In India the shortened form شهتوت is the common name of edible mulberries in general.

MACLURA

Maclura has only one species, of which there are specimens in the south-west corner of the Garden.

M. pomifera *Schneider*, The Osage Orange, Bois d'Arc, or Bow wood. It is native of the South and Central United States, where it is much used for hedges. The flowers are dioecious. The staminate flowers are arranged in loose racemes; the carpellary ones are in a dense, globose head which in fruit superficially resembles a large orange, but is not edible. This tree sometimes attains a height of 50–60 feet. The hard, strong, flexible wood is bright orange-coloured, but turns brown on exposure. It was formerly used by the Osage and other Indians west of the Mississippi for bows and war-clubs. It is now largely used for fence-posts, etc. The bark of the roots yields a yellow dye, and that of the trunk is sometimes used for tanning.

DORSTENIA

Dorstenia has about 70 species, native of the tropics. There is only one species in Asia. Our specimens are in

MORACEAE

the Stove. The inflorescence, owing to the horizontal expansion of the axis, is usually more or less disc-like.

D. contrajerva *L.*, is native of tropical America, where it is used medicinally.

FICUS (FIG)

Ficus has about 600 species. Nearly all of them are native of the tropics. Note the large stipules, which envelope the buds, and drop off as the leaves expand. Our specimens will be found in the Stove and Palm House. *Ficus carica* grows on the wall of the laboratory.

F. carica *L.*, The Common Fig. Native probably of West Asia and of the eastern and southern Mediterranean Region. It is said to have been introduced into South Italy by Greek colonists, and has been cultivated from early times in many lands.

The fig is a multiple fruit. Such a fruit is derivable from the flat, disc-like inflorescence of *Dorstenia* by drawing the edges of the disc together so as to make them nearly meet, and thus forming a fleshy bag, whose inner surface is lined with small fruits and whose aperture is almost closed. For an account of the pollination see Müller's *Fertilisation of Flowers*, p. 521.

The fig is constantly mentioned in oriental literature. In Arabic it is called تين (Hebrew תְּאֵנָה in the Old Testament), and in Persian انجير, a word widely used in the East. In the Sura وَالتِّين, an oath is made by the Fig, the Olive, and Mount Sinai.

In the description of the garden in the first book of the 'Lights of Canopus' the flavour of the fig is aptly compared to a mixture of opium and sugar:—

در یك جانب انجیر بینظیر که دست قدرت وصف جمالش بر
طبق وآلتّین نهاده حلوای زیبا از خشخاش و قند ترتیب داده

F. sycomorus *L.*, Sycamore, Fig Mulberry. Native of North Africa and Palestine. This tree is the Sycomore of the Bible and the συκάμινος ἡ Αἰγυπτία of Theophrastus, who describes

MORACEAE

its habit of bearing fruit on its trunk (*Hist. Plant.* 1, i, 7). Some writers use the spelling sycomore for this tree and sycamore for *Acer pseudoplatanus*.

The Hebrew word שִׁקְמָה occurs frequently in the Old Testament. The Greek word in the New Testament translated sycamore is συκόμορος ('Fig Mulberry'), a word doubtless derived from the Semitic by assimilation with σῦκον ('fig') and μόρον ('mulberry'). The ancient Egyptian mummy-cases were made of the wood of this tree.

F. bengalensis *L.*, Banyan. Native of India and Africa. The Banyan in its native land often attains a height of 100 feet. Its huge, horizontal branches throw down at intervals aerial roots which thicken into pillar-like trunks. The Hindus venerate this tree and encourage the growth of the roots by protecting them with bamboo stems. The seeds often germinate in other trees, which are eventually strangled by the growing roots. In India the Banyan is called بَر and بَرْگَت.

F. religiosa *L.*, Bo-tree (of Buddha). Probably wild in the Siwaliks and in the forests of the Himalayan foot-hills. Hindus and Buddhists, who hold this tree sacred, plant it in all parts of India, Ceylon and Burma. The written records of the Bo-tree at Anuradhapura in Ceylon go back more than 2000 years. Like the Banyan it usually starts life as an epiphyte. The drip-tips of the leaves are well known (see Schimper, p. 19). In the Sanskrit classics it is called पिप्पल and अश्वत्थ; the first of these names is common in the vernaculars (e.g. پیپل).

F. elastica, India-rubber Plant. Native in the outer Himalaya from Nepal to Assam, in the Khasia Hills, and Burma. It sometimes reaches 200 feet in height. This is one of the trees that yields rubber or caoutchouc, which is obtained by agglutinating the milky juice which flows freely when any part of these plants is cut or broken. The word rubber is sometimes used to denote the crude product, and caoutchouc the pure hydrocarbon isolated from it. Rubber-

MORACEAE

yielding trees belong to the *Moraceae*, *Euphorbiaceae* and *Apocynaceae*.

Gutta-percha, which differs from rubber in softening and becoming plastic in hot water, is obtained chiefly from trees belonging to the *Sapotaceae*.

Family NYMPHAEACEAE

NELUMBIUM

Nelumbium has 2 species; *N. lutea* Willd., native of Atlantic North America, and the Sacred Lotus. The carpels are embedded separately in the top of the obconical receptacle. The Sacred Lotus grows in the Tropical Aquarium.

N. speciosum *Willd.*, Sacred Lotus. Theophrastus (*Hist. Plant.* IV, viii, 7 and 8) gives a beautiful and accurate description of this plant, the Sacred Bean of the ancient Egyptians. Though it is now widely distributed in Asia and found in parts of North Africa and tropical Australia, it no longer grows in Egypt. It is common throughout India, where the Hindus, who call it कंवल, have venerated it for countless generations. In Islamic literature it is confounded with other Water-lilies under the name نيلوفر (a word derived from Sanskrit नीलोत्पल, 'blue water-lily' from नील, blue, and उत्पल, water-lily). Water will not wet the leaves but runs on their surface like quicksilver. The rootstock and seeds are edible. The huge, rose-coloured flowers are of great beauty.

Family BERBERIDACEAE

BERBERIS

Berberis has about 60 species, native of the North Temperate Regions, and the Andes. The wood of all of them is yellow from the presence of an alkaloid called *berberin*, which is present in other genera of *Berberidaceae* as well as in the 'Calumba Root' of the British Pharmacopoeia

BERBERIDACEAE

(*Jateorhiza columba* Miers, Family *Menispermaceae*), and some other medicinal plants. Preparations of various plants containing berberin are used in native medicine for diseases of the eye in very diverse parts of the world. The species with pinnate leaves are sometimes called *Mahonia*, and those with simple leaves *Euberberis*. The species of *Euberberis* have long and short shoots. The leaves of the long shoots are usually transformed into tripartite spines. In their axils are the short shoots, which bear green leaves and racemose inflorescences. The stamens of all the species are sensitive, bending inwards when their inner surface is touched. *Berberis vulgaris* L., the Common Barberry, the only British species, bears the *aecidium* stage of the fungus which causes the 'rust' of wheat and other cereals. *B. aquifolium* Pursh., the Oregon Grape, a western North American species with pinnate leaves, is naturalised in Britain and grows on Fleam Dyke. Our collection is to the south of the Lynch Walk.

× **B. stenophylla** *Moore*. This is a hybrid between two Chilean species, *B. darwinii* Hooker and *B. empetrifolia* Lam. Specimens of both parent plants will be found near the *B. stenophylla* in the collection close to the Lynch Walk. This is one of the loveliest of flowering shrubs. In May our specimen opposite the west entrance of the Houses looks like a fountain of gold.

B. fremontii *Torrey*. Native of the south-western United States. The pinnate leaves of this species bear from five to seven small glaucous leaflets, whose margins are beautifully scalloped between the slender spines. Our plant, which will be found in Bay No. 2, is probably among the best specimens in this country.

B. fortunei *Lindl*. Native of China. A specimen of this species will be found beside *B. fremontii* in Bay No. 2. The narrow leaflets, which taper toward both ends, are very different from those of any other species.

PLATE XIII

ASIMINA TRILOBA IN FLOWER, BETWEEN THE STREAM AND THE POND (p. 55)

PLATE XIV

AN AUSTRALIAN SUNDEW (*DROSERA BINATA*), IN THE
TROPICAL AQUARIUM (p. 58)

MENISPERMACEAE

Family MENISPERMACEAE

Most of the species of this family are tropical. Two North American species, whose ranges extend far northwards, will be found climbing on the west wall of the side entrance. The leaf-stalk of *Menispermum canadense* L. (Moonseed) is attached to near the edge of the large, 3–7-angled or -lobed leaf-blades. In *Cocculus carolinus* DC. the stalk is attached to the edge of the ovate or cordate, much smaller, almost entire blade.

Family MAGNOLIACEAE

MAGNOLIA

Magnolia has about 30 species, native of tropical Asia, East Asia, and Atlantic North America. The fossil remains of numerous species occur in the Cretaceous and Tertiary deposits of Europe and Greenland. Note on the twigs the large, nearly horizontal, shelf-like leaf-scars, and the ring-like scars of the stipules. The leaves are always entire. Our collection is between the west end of the Border and the Pterocarya thicket.

M. grandiflora *L.*, Laurel Magnolia, Bull Bay. Native of Atlantic North America, where it sometimes attains a height of 80 feet. In this country it has been grown from early in the eighteenth century, and seldom reaches half that height. The leaves are evergreen and of the Laurel type. They resemble those of the Cherry Laurel (*Prunus laurocerasus* L.), but differ from them in being rusty-brown beneath. (The only other hardy evergreen *Magnolia* is *M. delavayi* Franch.[1], whose leaves are glaucous beneath.) The huge, globular, deliciously fragrant flowers appear in this country usually towards autumn, but in America they bloom from April to August. The timber is little used except for fuel.

M. acuminata *L.*, Cucumber Tree, Mountain Magnolia. Native of Atlantic North America, where it reaches a height of 90 feet. It was introduced into England in 1736. It grows with us into a much larger tree than *M. grandiflora*.

[1] This species is hardy only in the south-western counties. Our specimen is in the Temperate House.

The greenish, bell-shaped flowers, which appear during the summer amongst the foliage, are comparatively inconspicuous. The unripe fruits bear a slight resemblance to Cucumbers; hence the name Cucumber Tree. The timber of this species is sometimes used for flooring and cabinet-making. Our specimen stands by the hedge, next to the Tulip Tree.

M. stellata *Maxim.* Native of Japan. It was introduced into this country in 1877. This species is a compact shrub with bitter, aromatic bark. The flowers appear in early spring before the leaves. The petals, which are more numerous and narrower than those of other Magnolias, spread out to form a beautiful white star.

M. conspicua *Salisb.*, Yulan, Lily Tree. Usually said to be native of China, but it was not known to the ancient Chinese, and its common Chinese name, *Hsin-i*, means 'bitter barbarian.' In Chinese literature it is first mentioned by the poet Wang Wei, who flourished in the first half of the eighth century A.D. The large, white flowers appear in the spring before the leaves and are thought by the Chinese to resemble jade. *Yü-lan* means jade 'lan' (*lan* being the name of plants whose flowers have a similar odour[1]). The men of the Sung dynasty (960–1260 A.D.) named the plant *Ying ch'un*, which means 'meet the Spring.'

LIRIODENDRON

Liriodendron has 1 species, native of Atlantic North America and China. Some consider the Chinese form to be a distinct species. Remains of allied trees are found in the Tertiary deposits of Europe and Greenland. Our specimen of the Tulip Tree is by the hedge behind the Magnolias.

L. tulipifera *L.*, Tulip Tree, Yellow Poplar. Native of Atlantic North America. In its home it sometimes approaches 200 feet in height, with a trunk destitute of branches

[1] *Lan* is given in Professor Giles's *Dictionary* as "a general term for orchidaceous plants." Among others there is the *mu lan* (wood or wooden lan), *Magnolia obovata* Thunb., which is mentioned by the poet Ch'ü Yüan, B.C. 332–295.

MAGNOLIACEAE

for 80 to 100 feet from the ground. The leaves are broad and have two basal lobes. The apex of the leaf is so broadly and abruptly truncate that it appears as it were cut off with a pair of scissors. It is to the beautiful tulip-like flowers that the tree owes its popular and its trivial name (λείριον, lily, and δένδρον, tree). The wood is used for various purposes. The inner bark, especially of the root, is used in domestic medicine.

DRIMYS

Drimys has 10 species, most of them are natives of the southern hemisphere. Several are used medicinally. Our specimen of Winter's Bark is in Bay No. 5.

D. winteri *Forster*, Winter's Bark. Native from Mexico to Terra del Fuego. Captain Winter brought home the bark of this tree in 1578, and for a long time it was much used in medicine, being considered a sovereign remedy for vomiting, scurvy, and palsy. It has now died out of use. The plant was introduced into this country in 1827.

Family ANONACEAE

ASIMINA

Asimina has about 6 species, natives of the eastern United States. Our best specimen of the Papaw is near where the stream enters the Water Garden.

A. triloba *Dunal.*, Papaw (Pls. XII and XIII). Native of eastern North America, where it extends northward to Ontario, Erie, and Michigan, being the only representative of the family *Anonaceae* whose range extends far outside the tropics. In the western part of the prairie region it grows in the forests that fringe the rivers as far north-west as eastern Nebraska. It is very common in the Mississippi valley, forming thick undergrowth on rich bottom-lands, or thickets many acres in extent. The buds are naked, but are protected by a rusty down. The fruit is sweet, luscious, and wholesome. The name Papaw is also used for another tree, *Carica papaya* L. (Fam. *Caricaceae*), which is cultivated throughout the tropics for its melon-like fruit.

Family LAURACEAE

Umbellularia

There is only one species of this genus. Our specimen is in Bay No. 7.

U. californica *Nutt.*, Californian Laurel. The figure of a twig of this plant on p. 513 of Schimper is familiar to many.

It is an abundant and characteristic tree of California, specially common in moist places. It attains its greatest height (90 feet) in the cool, fog-moistened valleys of the coast of North California and South Oregon. It often crowns the highest points of the coast-range hills, up to 2500 feet. Together with *Acer macrophyllum* Pursh., it forms much of the forest growth in south-western Oregon. The leaves abound in a volatile oil which causes them to burn vigorously, even when green. Sniffing the crushed leaves too freely may cause sneezing and even head-ache.

Laurus

Laurus has 2 species, one native of the Mediterranean Region, the other of the Canary Islands and Madeira.

L. nobilis *L.*, Bay Laurel. Native of the Mediterranean Region, where it is a relict of the Tertiary flora. Many plants are popularly known as 'Laurels,' but the Bay Laurel has the best claim to the title as it is the '*laurus*' of the ancients, sacred to Apollo, and the symbol of victory ('At te uictrices lauros, Messalla, gerentem portabat nitidis currus eburnus equis,' Tib.). According to Suetonius, Tiberius, during cloudy weather, wore a laurel crown on his head to protect himself from lightning. He must have supposed that Apollo would be unwilling to strike his own tree. Nowadays the leaves are used for flavouring milk puddings. In Persian the Laurel is called درخت غار. Cataplasms of the leaves are supposed in the East to be good for bee- and wasp-stings.

Family CAPPARIDACEAE

CAPPARIS

Capparis has about 150 species, native of the warmer parts of the world. Our specimen of the Common Caper will be found in Bay No. 2.

C. spinosa *L.*, The Common Caper. Native of the Mediterranean Region and very abundant in walls and rocky clefts in the south of Europe and North Africa. The stipules are represented by little spines. The internode between the androecium and gynaeceum is long and forms the stalk (gynophore) which bears the ovary, and later the fruit. 'Capers' are the unexpanded flower-buds of this species. In the East they are called كبر.

Family SARRACENIACEAE

The three families *Sarraceniaceae*, *Nepenthaceae*, and *Droseraceae*, all of whose members have insect-catching leaves, make up the Order *Sarraceniales*. Most of our insectivorous plants will be found in the Tropical Aquarium.

The nine members of the *Sarraceniaceae* are native of the New World. They are perennial herbs, whose pitcher-like leaves are arranged in rosettes. The mouth of each pitcher is thickened to form part of a ring, which is usually surmounted by a kind of lid. The ventral side of the pitcher is drawn out into a keel or wing. The new pitchers appear in the spring, but in some species another set of them arise in the autumn, while yet other species at that season produce flat green leaves.

The mechanism for entrapping insects varies somewhat in the different genera. In *Sarracenia* honey-glands are scattered over the outside of the pitcher and the inner surface of the lid is thickly beset both with honey-glands, and stiff hairs which point downwards. The ring and upper part of the inside wall form a slippery zone, below which the wall of the pitcher bears downwardly-directed hairs like those on the lid. It is easy to understand the passage of an insect from the honeyed regions to its fall into the sticky fluid at the bottom of the pitcher, where it is drowned, but not digested. When the pitchers rot away in the autumn the disintegrated remains of the insects supply the roots with valuable nitrogenous matter. The names of the three genera are *Heliamphora*, *Sarracenia*, and *Darlingtonia*. Note the curious umbrella-like style of *Sarracenia*; the small stigmas are on its lower surface.

Family NEPENTHACEAE

This family has only one genus, *Nepenthes*, the greater part of whose species grow in the Indo-Malay Region. They are scrambling shrubs with alternate leaves, which end in tendrils. The end of the tendril of the lower leaves develops a pitcher very much like that of the *Sarraceniaceae*. The juice which these pitchers contain, unlike that of the *Sarraceniaceae*, resembles gastric juice and can digest large insects, whose products are absorbed by the leaf. Neither are the pitchers of the two families homologous. Those of the *Sarraceniaceae* represent entire leaves, whereas in *Nepenthes* the proximal part of the leaf forms a blade, and the distal, tendrillar part develops the pitcher.

Family DROSERACEAE

In this family the leaves, though not pitcher-like, are yet able to catch and digest insects.

DROSERA (SUNDEW)

Drosera (Pl. XIV) has 90 species, 3 of which are British. The leaves are beset with tentacle-like hairs, which secrete a sticky substance, and which, when insects become entangled among them, bend inwards. The substance then becomes acid and digestive. The other three genera of Droseraceae contain one species each:—

Drosophyllum lusitanicum *Link*. Native of South Spain, Portugal and Marocco. The long, narrow leaves bear two

DROSERACEAE

kinds of glands; stalked ones, which secrete a sticky substance, and sessile ones, which, when nitrogenous matter comes into contact with them, secrete a digestive fluid. Insects that become entangled by the stalked glands sink dead into the sessile ones which digest them.

Dionaea muscipula *Ellis*, Venus's Fly-trap, is native of a very small area of land near Wilmington in North Carolina. Each leaf consists of a flat stalk, and a two-lobed blade, which opens and shuts like a book. Each lobe of the blade bears on its margins numerous coarse bristles, which interlock when the 'book' closes, and on its inner surface three very fine bristles which are sensitive. When the plant grows in sunshine the inner surface of the lobes develops red glands, which are attractive to insects. A single contact with one of the sensitive bristles has no effect, but when two contacts are made at a short interval with one bristle, or, when two of the bristles are touched within a short interval, the blade shuts firmly. When an insect is caught an acid, digestive fluid is poured out upon it after a few hours, and the leaf ultimately absorbs its dissolved tissues.

Aldrovanda vesiculosa *L.* has a scattered distribution from Queensland to Central Europe. It is a free-floating water plant whose leaves are arranged in whorls of about eight. They somewhat resemble the leaves of *Dionaea*, but the flat stalk ends in about half a dozen bristles which are longer than the book-like blade. The 'book' bears sensitive bristles, star-like hairs, and papillae. It does not shut up flat like that of *Dionaea*, but, when the edges meet, forms a lenticular chamber.

Family CEPHALOTACEAE

Cephalotus follicularis *Labill.*, the only member of the family, grows in swamps in Western Australia. Its leaves are of two kinds, ordinary flat leaves, and pitcher-leaves which are very like those of *Nepenthes*. The pitchers secrete a digestive fluid. This family is not included in the Order *Sarraceniales*.

Family SAXIFRAGACEAE

ESCALLONIA

Escallonia has about 50 species, native of South America. They are very abundant in the sclerophyllous vegetation of Chile. The twigs and leaves are often dotted with resinous glands which give the plant a strange aroma. The scent of the Chilean vegetation is said to extend several miles out to sea. The smell of *E. illinita* Presl., and *E. viscosa* Forbes, may be likened at will to pig-sties or to liquorice. *E. macrantha* Hooker is commonly planted in the south-west of England in garden hedges. Some say that the leaves of this species are good and wholesome to eat, but one who partook of them complained bitterly of their effects. Our collection of Escallonias occupies a bed on the inner side of the Walk near the north-west corner of the Garden.

Family HAMAMELIDACEAE

A collection of plants of this family will be found at the south end of the Middle Walk.

LIQUIDAMBAR

Liquidambar has about 4 species which are distributed discontinuously in different parts of the world. These plants are popularly confused with Maples, from which they differ in their alternate leaves, and with Planes, from which their inconspicuous stipules at once distinguish them. The flowers are naked, and arranged in dense, usually globular, heads.

L. orientale *Miller*. Native of Asia Minor. 'Styrax preparatus,' or Prepared Storax, is defined in the British Pharmacopoeia as "a viscid balsam obtained from the wounded trunk of *Liquidambar orientalis* Mill., purified by solution in alcohol, filtration, and evaporation of the solvent." This substance is contained in Friar's Balsam (Tinctura Benzoini Composita). The سلارس (lit. rock exudation) of Indian bazaars is either the balsam, or the oil

PLATE XV

THE MUSK ROSE (*ROSA MOSCHATA*), CLIMBING UP THE
AUSTRIAN PINES ON THE HILL (p. 63)

PLATE XVI

THE JUDAS TREE (*CERCIS SILIQUASTRUM*) (p. 68)

HAMAMELIDACEAE

expressed from the seeds, of this tree. (سلاجت is probably always a mineral product.) In Arabic and Persian the balsam is called ليٮنى and مياه سائلم ('running waters').

L. styraciflua *L.* is native of the eastern United States, where it surpasses all other trees in the brilliance of its autumn colours. A specimen will be found in the Collection.

Parrotia

Parrotia has 1 species, of which we have a specimen by the Judas Tree.

P. persica *C.A. Meyer*, Iron Wood. Native of North Persia and the Caucasus, where it forms tangled thickets. The branches, and even the main stems, where they touch each other, grow together by a process of natural grafting. Note the flaking, plane-like bark. The flowers, which appear in the spring, have no petals, but the bright red stamens are very conspicuous. The timber is red and hard, hence the Tartar name دمير اغاجى, from دمير (تيمور), iron, and اغاج, tree.

Hamamelis (Witch-hazel)

Hamamelis has 4 known species, one is native of eastern North America, and the others of Eastern Asia. The Witch-hazels have both sepals and petals. The petals are remarkably long in proportion to their breadth. Specimens of all 4 species will be found in the collection near the south end of the Middle Walk.

H. virginiana *L.*, Witch-hazel. Native of eastern North America, introduced into this country in 1736. The early settlers named the plant after the resemblance of its leaves to the Hazel. 'Witch' is perhaps for 'Wych' from O.E. *wice*, the name of a tree. This word probably originally meant pliable or drooping (cf. Eng. *weak*, Ger. *weich*, Da. *veg*). See Wych-elm, p. 46. The bark (Hamamelidis cortex) and the leaves (Hamamelidis folia) of this species are in the British Pharmacopoeia. Their valuable astringent properties are due to their richness in tannic acid, of which the bark contains as much as 8 per cent. *Liquor hammamelidis*, the so-called hazeline, is an extract distilled from the leaves.

EUCOMMIACEAE

Family EUCOMMIACEAE

Eucommia ulmoides *Oliver*. This tree, of which there is a specimen at the south end of the Broad Walk, is the only member of the family *Eucommiaceae*. It has never been found wild. The Chinese, who value its bark as a tonic, have cultivated it for thousands of years. The plant abounds in latex, which yields a kind of gutta-percha.

Family PLATANACEAE

PLATANUS (PLANE)

Platanus, the only genus, has about 6 species, native of the North Temperate Regions. The Planes are trees with scaling bark. The leaves are alternate, and the stipules are united to form a tube, which is crowned by a frill. Note how the enlarged base of the petiole encloses the axillary bud. The globular heads of flowers are well known (cf. *Liquidambar*, p. 60, and *Acer*, p. 79). Our Planes are on the east side of the East Walk.

P. orientalis *L.*, Oriental Plane. Native of South-east Europe and Asia Minor, and perhaps from thence to the Western Himalaya. In this species each leaf has five to seven very narrow lobes. In England it has been cultivated from the sixteenth century, but the rapidly growing 'London Plane' (*P. acerifolia* Willd.) is now supplanting it. The Oriental Plane is the *Platanus* of Latin authors, and the چنار of the East. Persian poets liken its leaves to hands.

پنجه‌ها چون دست مردم سر فرو کرد از چنار—فرّخی

"*Human-like five-fingered hands reach downwards from the Platanus.*"

It is extensively planted as a shade tree in Mediterranean towns. The timber is tough and ornamental.

P. acerifolia *Willd.*, The London Plane. This is by far the commonest tree in London streets. The leaves have three to five broad, usually triangular, lobes. This tree has never been found wild. Some regard it as a variety of

PLATANACEAE 63

P. orientalis L., others as a hybrid between that species and *P. occidentalis* L., an American species which is not hardy in this country.

Family ROSACEAE

Pyrus (Apple, Pear)

Pyrus has 50 to 60 species, native of the northern temperate and cold regions. Our collection is in the south-east corner of the Garden.

P. salicifolia *Pall.*, Willow-leaved Pear. Native of the Caucasus. The flowering of this tree is one of our most striking annual occurrences. The masses of pure white flowers and silky white leaves, which resemble those of the White Willow, usually appear together in April, clothing the tree in a mantle of white splendour.

Rosa (Rose)

Rosa has about 100 known species, nearly all native of the North Temperate Regions. Our collection is between the Middle Walk and the Pond.

R. macrophylla *L.* Native of the Himalayas and West China. This species is remarkable for its stout, tree-like stems, and large leaves. Our specimen, which stands on the inner side of the path at the north-west corner of the Garden, has very few thorns, and some specimens are quite thornless.

R. moschata *Miller*, Musk Rose (Pl. XV). Native from South-east Europe to North India and China. With us this Rose climbs up the Austrian Pines on the Hill producing among their crowns its huge clusters of white flowers.

Prunus

Prunus has 150 species, most of them native of the North Temperate Regions. Five are British (sloe, bullace, gean, bird-cherry and true cherry, see *British Floras*). Note that the leaves are folded differently in the buds of the different species. Our collection is near the Houses on the west side of the Middle Walk.

P. armeniaca *L.*, Apricot. Native certainly of China, and

ROSACEAE

perhaps also of other parts of Asia. The Apricot is a true species, and not as is popularly supposed a hybrid between the peach and the plum, though it shares some of the features of both these species. It has been cultivated for its fruit from early ages. The תַּפּוּחַ of the Canticles, translated 'Apple' is probably the Apricot; see Tristram, Art. *Apple*, p. 334.

P. amygdalus *Stokes*, Almond. Both sweet and bitter varieties of Almond have been long cultivated in the Mediterranean Region, in parts of which the tree is probably native. The fruit, like that of all the other species of *Prunus*, is a drupe. The familiar Almond is the seed. The stone which contains it is finely sculptured with lines and pits. In the edible almonds the stones are not nearly so hard as those of the cherry and plum, but the varieties commonly cultivated for their flowers have very hard stones. All know the beauty of the flowers of the Almond borne in early spring on leafless branches. The Hebrew name שָׁקֵד means watching or waking (i.e. from winter sleep). לוּז, in Gen. xxx, 37 translated Hazel, is probably the Almond (cf. Arabic لوز). See Tristram, Art. *Almond*, p. 332. The Persian name بادام has spread widely with Islamic culture.

P. triloba *Lindley*, native of China, is allied to the Almond. There is a specimen of an exquisite double-flowered variety on the east wall of the Side Entrance.

P. mahaleb *L.*, St Lucie Cherry. Native of Central and South Europe and of parts of Asia. All parts of this tree have, especially when dry, a pleasant aroma of *coumarin*, a substance present also in hay, woodruff, and tonka beans. The 'Weichselrohr' or cherry-wood pipe-tube, now rarely seen in this country, is made of its wood; but the cherry-wood pipe-bowl is usually made of the wood of other species of *Prunus*. The type raised from seed is used as a stock for grafting cherries. The trivial name is from the Arabic محلب. In the East they use the stones of the fruits in medicine and perfumery. In Persia, where the tree is called پیوند مریم, a cataplasm of the seeds is applied to bruises.

ROSACEAE

P. laurocerasus *L.*, Cherry Laurel. Native of the Eastern Mediterranean Region. The leaves of this species are evergreen and of the laurel-type (see note, p. xv). The flowers are arranged in racemes and the cherry-like fruits are fairly good to eat. 'Laurocerasi folia' of the British Pharmacopoeia are the fresh leaves of this species. They contain hydrocyanic acid, and should not be substituted for those of *Laurus nobilis* as a flavouring.

P. lusitanica *L.*, Portugal Laurel. Native of Spain and Portugal, Madeira, the Canaries, and Azores. This species is allied to the Cherry Laurel, but has smaller leaves which contain no hydrocyanic acid, and longer racemes. Compare the odour of the crushed leaves of these two species. In Killarney this species seeds itself and is becoming wild.

NUTTALLIA (*OSMARONIA*)

N. cerasiformis *Torrey*, the Oso Berry, is the only species. It is native from British Columbia to California. Whereas in the genus *Prunus* there is only one carpel, which becomes a drupe, in *Nuttallia* there are five carpels, of which sometimes all, but rarely more than two or three, develop into drupes. Our specimen is in the *Prunus* Collection.

Family LEGUMINOSAE

Our collection of hardy leguminous trees and shrubs is on the Cornel Plot.

ACACIA

Acacia has about 500 species; nearly all are natives of the tropics. Their leaves are either bi-pinnate, or else consist entirely of a flattened stalk called a *phyllode*. That these phyllodes are leaf-structures and not stem-structures (see *Phyllocladus*, p. 3) is shown by their having buds in their axils; and the development of the seedlings makes it clear that they represent the leaf-stalks alone. The first foliage-leaves of the seedling are simply pinnate, and the succeeding leaves bi-pinnate. Then follow leaves with reduced leaf-blades and somewhat flattened stalks. Finally the blades completely disappear and the flattened stalks (phyllodes) alone remain. Sometimes the phyllodes of adult plants bear bi-pinnate blades. The 280 phyllode-bearing Acacias are native of South and West Australia, where they are an important constituent of the 'scrub,' and also of the Pacific islands.

Specimens of phyllode-bearing Acacias will be found in the Temperate House. Among them are:—

A. armata *R. Br.*, The Kangaroo Thorn, which makes a good hedge plant in countries where it is hardy. The phyllodes of this species are short, sharp-pointed, and traversed by a midrib. The edge of the phyllode next the stem is evenly curved, and the outer edge wavy. The stipules are spinous.

A. pycnantha *Benth.*, The Golden Wattle. The phyllodes are broad and leaf-like, and sometimes as much as six inches long. This is one of the species the barks of which are highly valued for tanning. Indeed its bark is said to contain a higher percentage of tannin than that of any other species, but it does not grow into such a big tree as *A. decurrens* Willd., the Black Wattle, which is the best-known tan-yielder.

Of Acacias with bi-pinnate leaves we have several specimens:—

A. dealbata *Link.*, The Silver Wattle. Native of South Australia. It is cultivated in the south of Europe, and early in the spring flowering branches of this and other species are sold in the London streets as 'mimosa.' (In the genus *Mimosa*, which was formerly included in *Acacia*, the stamens are 5 or 10 in number; in *Acacia* they are numerous.) This species is often considered to be a variety of *A. decurrens* Willd., but differs from that species by its silver-grey foliage and by broader pods not contracted between the seeds. Our specimen is in the Temperate House.

A. spadicigera *Cham.* Native of Jamaica. This is one of the 'myrmecophilous' Acacias mentioned by Schimper (p. 140). The stipules of these plants are represented by large, hollow thorns into which ants bore their way and make

LEGUMINOSAE

their dwelling. The small yellow pellets at the ends of some of the leaflets are the food of these thorn-dwelling ants which, in return for the shelter and sustenance they enjoy, protect the Acacia against the inroads of leaf-cutting ants. Our specimens are in the Stove.

COPAIFERA

Copaifera has 24 species, of which 16 are native of tropical America, and 8 of Africa. Our specimens are in the Stove. 'Copaiba' of the British Pharmacopoeia is the oleo-resin obtained by incision from the trunks of various species of Copaifera. It comes from the hotter parts of South America. The stems of the trees which yield it are often 6 feet thick, and are traversed by canals nearly an inch in diameter, which contain a great abundance of oleo-resin. The stems sometimes burst spontaneously with a loud report, liberating as much as 50 kilos of oleo-resin. When the trees are incised the oleo-resin is collected in lead vessels of 22 litres capacity, into which it runs in a viscid stream dulled by air-bubbles. The flow sometimes ceases for several minutes and then a gurgling noise announces its speedy resumption. Some of the African species yield hard resins known as copals, which are used for making varnish.

BAUHINIA

Bauhinia has 200 species, nearly all are natives of the tropics. As the leaves of most of the species are more or less deeply cleft from the tip into twin lobes, or leaflets, suggesting two brothers, it was thought fit to name the genus after John and Caspar Bauhin, sixteenth-century herbalists.

B. yunnanensis *Franch*. Native of western China. This species is not usually considered hardy in Great Britain, but our plant, which is in Bay No. 4, survives in the open air because it is sheltered on all sides but the south, and derives warmth from the wall of the Palm House. It flowers in July. The flattened tendrils arise in the axils of scale-leaves, and are therefore branches. The leaves of this species are completely divided into two leaflets which rest at night folded like the wings of a butterfly.

B. variegata *L.* Native of India, Burma, and China. This species is a tree, and is the well-known كچنار of India, where it is much planted. The flowers which are a glorious medley of red, cream, and purple, appear in the spring before the new foliage. All parts of this plant are used medicinally. The flower-buds are eaten as a vegetable, and the bark yields dye, tan, and fibre, as well as medicine. The gum, in which this and several other Indian species of *Bauhinia* abound, is locally used for various purposes, but is of no commercial importance. Our specimen is in the Palm House.

Cercis (Judas Tree)

Cercis has 7 species, native of South Europe, East Asia, and North America. Our collection is near the south end of the Middle Walk.

C. siliquastrum *L.*, Judas Tree (Pl. XVI). Native of the Mediterranean Region. According to legends Judas Iscariot hanged himself either on a tree of this species, or on an Elder. It is the ارغوان of the Persians, in whose poetry it is often mentioned on account of its fine rose-purple flowers.

ارغوان لعل بدخشی دارد اندر گوشوار—فرّخی

"*Ruby ear-rings of Badakhshán sparkle on the Judas Tree.*"

In Islamic literature a beautiful complexion is frequently likened to the flowers of this tree, and often contrasted with saffron. In the Urdu *Arabian Nights* we read

روز بروز چهره ارغوانی اسکا زعفرانی هوتا جاتا تها

"*Day by day his Judas-flower complexion gradually became more and more saffron-coloured.*"

The flowers, which appear before the leaves on the boughs and twigs, look papilionaceous because the two lower petals form a kind of keel which encloses the stamens and carpel, but note that the large upper petal in bud is within the two side petals, an arrangement characteristic of the sub-family

PLATE XVII

THE HEDGEHOG BROOM (*ERINACEA PUNGENS*), IN BAY No. 2 (p. 70)

PLATE XVIII

THE TRIFOLIATE ORANGE (*AEGLE SEPIARIA*), BETWEEN
BAYS Nos. 5 AND 6 (p. 75)

LEGUMINOSAE 69

Caesalpinioideae. In the sub-family *Papilionatae* (pea-flowers) the upper petal, which is called the *vexillum* or standard, is, in bud, exterior to the side petals, and folds over them.

Our specimen, a particularly fine one, is remarkable in having no main trunk. About a dozen stems spring from the ground and most of these are spirally twisted and spread for a long distance horizontally.

GYMNOCLADUS

Gymnocladus has 2 species, one native of eastern North America, the other of southern China. Our specimen of the Kentucky Coffee-tree is on the east of the Middle Walk near where it joins the South Walk.

G. dioica *Koch* (*G. canadensis* Lamk.), The Kentucky Coffee-tree. Native of eastern and central United States. The early settlers in Kentucky used the seeds as a substitute for coffee. Note the long, stout, bare branches from which is derived the generic name *Gymnocladus* (γυμνός, naked, and κλάδος, branch). In autumn the leaflets fall from the leaf-shafts which remain for some time attached to the branches. The bark contains saponin, a substance which froths in water.

GLEDITSCHIA

Gleditschia has 12 species, very widely distributed. The formidable, often branched, thorns are developed from the uppermost of a row of serial buds. Our collection is on the east of the Middle Walk.

G. triacanthos *L.*, Honey Locust. Native of eastern North America. In the valleys of small streams in southern Indiana and Illinois it sometimes reaches a height of 140 feet, with a trunk occasionally 5 to 8 feet in diameter. Our specimen of the Honey Locust suffered great damage in a storm a few years ago. We have excellent specimens of several other species.

CLADRASTIS

Cladrastis has 3 known species. Two are natives of China and the other of North America. Our specimen of the

Yellow-wood is on the east of the Middle Walk near its south end, behind the Kentucky Coffee-tree.

C. lutea *Koch* (*C. tinctoria* Raf.), Yellow-wood. Native of the south-eastern United States. The wood yields a yellow dye. The petiole enlarges abruptly at the base into a hollow cone which encloses several serial buds. Compare this with the similar hollow cone of the Plane Trees (*Platanus*), which contains only one bud. In two tribes of *Papilionatae* the stamens are free: the *Sophoreae* (leaves pinnate) and the *Podalyrieae* (leaves simple or palmate). *Cladrastis* belongs to the *Sophoreae* and differs from *Sophora* by its flat pods.

Erinacea

E. pungens *Boiss.*, Hedgehog Broom (Pl. XVII), the only member of the genus, is native of the Mediterranean Region, and especially abundant in Spain, where it grows so plentifully on the mountains that horses can scarcely make their way through it. Our specimen in Bay No. 2 is perhaps the finest in the country. There is a smaller specimen in the Rock Garden. In April and May the plant is covered with violet-blue flowers. Clusius described and figured this shrub in his *Hispanica*, which was published in 1576. His description begins: " plane nova et tota elegans est haec planta." The small scale-like leaves are opposite, and the branches which arise in their axils grow into sharp spines. Very few *Leguminosae* have opposite leaves, but they may be seen in two species of *Genista* which grow in the Rock Garden. These are *G. horrida* DC., whose branches are hard, sharp thorns (Lat. *horridus*, standing on end, prickly), and *G. radiata* Scop., which has pliant and harmless branches.

Robinia

Robinia has about 8 species, native of North America, extending to Mexico.

R. pseudacacia *L.*, Locust Acacia. Native of the eastern United States. It was introduced into Europe by the younger John Tradescant early in the seventeenth century.

LEGUMINOSAE 71

The popular name Acacia is very misleading, as apart from the stipular thorns it has little in common with the true Acacias. The timber is very durable in contact with the ground and is also much used in ship-building. The flowers are white, and deliciously scented.

R. hispida *L.*, The Rose-Acacia. Native of mountainous districts from Carolina to Georgia. The young twigs and leaf-stalks are beset with glandular bristles. The large, deep rose-coloured flowers are scentless. In the same district grows *R. viscosa* Vent., recognisable by its sticky twigs.

CARAGANA

Caragana has 20 species, most of them native of Central Asia. Note the absence of the terminal leaflet which is usually present in the Tribe *Galegeae* to which *Caragana* belongs. Note also the curious, scaly short shoots, which produce annually clusters of leaves and flowers.

C. arborescens *Lamarck*, Pea Tree. Native of Siberia. The tough bark can be made into ropes, and the seeds are good food for poultry.

C. frutescens *DC.* Native from the south of Russia to Japan. Ascherson and Graebner[1] derive the word *Caragana* from the Kirghiz name of this shrub which, they say, means 'black ear' or fox, applied because foxes abound in that region among the shrubs. The word in question is perhaps the Turkish قره قولاق (قره, black, and قولاق, ear) which is the name of *Felis caracal* L., the Caracal, generally supposed to be the Lynx of the ancients.

HALIMODENDRON

H. argenteum *DC.*, The Salt Tree, is the only member of its genus, which is closely allied to *Caragana*. It is native of the salt steppes of West Asia. A specimen will be found on the south side of the Lynch Walk.

ASTRAGALUS

Astragalus is one of the largest genera of flowering plants. It has about 1600 species. They are found chiefly in the

[1] *Flora des Nordostdeutschen Flachlandes*, p. 443.

72   LEGUMINOSAE

North Temperate Regions; 3 are British. Note the partition which almost divides the ripe pod into two compartments. There is a collection on the Herbaceous Beds.

A. tragacantha *L.* Native of the Mediterranean Region. This species is a low shrub. The leaf-shafts, which are fused with the stipules, persist after the leaflets fall, as an armour of formidable, densely-imbricated thorns. This species grows very slowly. Bean (1, 221) says that he has known a plant for 20 years which is still only 12 inches high. This is one of the species of *Astragalus* that yield 'Tragacanth,' which the British Pharmacopoeia defines as a gummy exudation obtained by incision from *A. gummifer* Labill., and some other species of Astragalus. This substance is known in commerce as Syrian Tragacanth, and in the East it is called كتيرا.

GLYCYRRHIZA

Glycyrrhiza has about 12 species which are widely distributed in the world. This genus is closely allied to *Astragalus*. The Spanish Liquorice is on the Medicinal Bed.

G. glabra *L.*, Spanish Liquorice. Native of the Mediterranean Region, extending to Hungary. It is extensively cultivated. Large quantities are used in making plug-tobacco, in brewing stout, and in medicine. 'Glycyrrhizae radix' or Liquorice Root of the British Pharmacopoeia is the peeled root and peeled subterranean stem of this and other species of *Glycyrrhiza*. Its preparations are demulcent, and disguise excellently the taste of many nauseous drugs. The liquorice powder of our childhood is still in the British Pharmacopoeia, where it is called *Pulvis Glycyrrhizae Compositus*. It owes its active properties to the senna leaves it contains, its grit to sulphur, and its aroma to Fennel fruit. It contains also Liquorice Root and sugar. The sticks of Liquorice sold in the shops are made of an aqueous extract. Pomfret-cakes are cakes of liquorice made at Pontefract (formerly called Pomfret) in Yorkshire. In Arabic Liquorice Root is called اصل السّوس and in Persian شيرين بيان. In India it is called ملهٹی, and بيخ حيات (root of life).

ERYTHROXYLACEAE

Family ERYTHROXYLACEAE

ERYTHROXYLUM

Erythroxylum has 90 species, natives of the warmer parts of both hemispheres. Our specimen of the Coca Plant is in the Stove.

E. coca *Lamarck*, Coca Plant. Native of South America. This plant is the source of the alkaloid Cocaine. The natives of Peru for countless generations have been in the habit of chewing the leaves in order to dispel fatigue, to gain a sense of pleasure, and to increase their capacity for work. The leaves are folded lengthwise in bud, and the impressions of this folding remain as two faint lines on either side of the midrib. The petals bear appendages.

Family RUTACEAE

All the plants belonging to this family contain volatile oils in intercellular spaces, which, when the leaves are held up to the light, appear as shining dots, and when they are crushed liberate the oil, from whose 'rutaceous' odour the family of the plant may usually be guessed.

XANTHOXYLUM

Xanthoxylum has 15 species, native of temperate East Asia and North America.

X. alatum *Roxb.* Native of the Himalayas, Khasia Hills, China, and Japan. This species is easily recognised by its winged leaf-shafts, and formidable flattened spines, attached to the branches by broad corky bases, which show striations due to annual growth. Each spine readily breaks off from its base, exposing a smooth oval facet. In India the branches are used for making walking-sticks; the pungent fruits for flavouring food, for medicine, and for purifying water; and the bark for intoxicating fish. The leaves of the Indian plant have 5–11 leaflets. The Chinese and Japanese plant, which has 3–5 leaflets, is sometimes considered to be a distinct species, *X. planispermum* Sieb. et Zucc. Our best specimen is on a bed near the western edge of the Pond.

RUTACEAE

X. fraxineum *Willd.*, Prickly Ash, Toothache Tree. Native of Atlantic North America. Our specimen grows where the stream comes out of its little tunnel under the path. In America the bark and fruits are used for toothache.

RUTA

Ruta has about 50 species, native of the Mediterranean Region and Asia. There is a specimen of Rue on the Medicinal Bed.

R. graveolens *L.*, Common Rue, Herb of Grace. Native of the south of Europe. The volatile oil in which this plant abounds has a remarkably strong aroma. Rue from classical times has been used as a medicinal herb and has been an object of much superstition. It is thought to be the πήγἄνον of Theophrastus. In Islamic medical works it is called سُداب.

The volatile oil distilled from the leaves, though no longer in the British Pharmacopoeia, is still used medicinally in this country. It has a lower specific gravity than any other known volatile oil. It is easily oxidised and will ignite when incorporated into a pill mass with potassium permanganate.

DICTAMNUS

D. albus *L.*, Burning Bush. The only species of the genus. Native of Europe and North Asia. The inflorescence is beset with numerous glands, which secrete so much volatile oil that on a warm, still evening the air around the plant becomes inflammable.

PHELLODENDRON

Phellodendron has 4 species, native of East Asia. The generic name (φελλός, cork, and δένδρον, tree) refers to the corky bark of the type species, *P. amurense* Ruprecht. *P. japonicum* Maxim., of which there is a fine specimen in a bed between the West Walk and the Pond, has smooth bark, and differs from *P. amurense* also by its broader leaflets, very downy beneath, and by its hairless winter buds.

RUTACEAE

PTELEA

Ptelea has 7 species, native of North America. Our specimens of the Hop-tree will be found near *Xanthoxylum fraxineum*, and on a bed between the West Walk and the Pond.

P. trifoliata *L.*, Hop-tree, Wafer Ash, Shrubby Trefoil. Native of eastern North America, where it generally grows on rocky slopes near the borders of the forest, often in the shade of other trees. Introduced into this country in 1704. The bark and leaves are bitter and aromatic. The fruits, which resemble those of the elms (πτελέα, elm), are occasionally employed as a substitute for hops in the production of home-brewed beer.

GLYCOSMIS

Glycosmis has 6 species, native of the Indo-Malay Region. Our specimen is in the Stove.

G. pentaphylla *Correa*. This shrub is very abundant in many parts of India, where it often forms undergrowth in the forest somewhat as hazel does in our oak-woods. It is called بن نميو. The twigs are sold in bundles in the bazaars for tooth picks. The fruits are eaten, but they are small and devoid of good flavour.

AEGLE

Aegle has 3 species, native of Asia.

A. sepiaria *DC.* (*Poncirus trifoliata* Raf.), the Trifoliate Orange (Pl. XVIII), of which there is a specimen between Bays Nos. 5 and 6, is a thorny deciduous tree native of North China. It is extensively used as a stock for oranges, and also as a hedge plant.

Family SIMARUBACEAE

AILANTHUS[1]

Ailanthus has about 7 species, native of temperate and tropical Asia.

---

[1] The generic name was spelt with an 'h' by Desfontaines, who described the genus in 1786. The spelling 'Ailantus' first occurs in 1825 (DC. Prod.). Though 'Ailanthus' is a hellenistic simulacrum, yet according to the law of priority this spelling must stand.

SIMARUBACEAE

Our specimens will be found to the south-east of the Pterocarya thicket.

A. glandulosa *Desf.*, Tree of Heaven (Pl. XIX). Native of China; introduced into this country in 1751. Though its leaves are pinnate and its fruits winged, yet this tree is not related to the Ash. Its winged fruits are derived each from a single carpel; each fruit of the Ash represents a whole ovary (2 carpels). Note the large glandular teeth near the base of the leaflets and compare the leaf arrangement with that of the Ash. The Chinese name of this tree is *ch'u*. 'Aylanto' is an Amboyna name for *A. moluccana* DC. The larva of a Chinese silk-moth (*Attacus cynthia* Drury) feeds on the foliage of the Tree of Heaven and of *A. vilmoriniana*.

A. vilmoriniana *Dode*. Native of China; introduced into this country in 1897. This species differs from *A. glandulosa* in the young branches being beset with spines, and the leaflets being downy.

Family BUXACEAE

BUXUS (BOX)

Buxus has 20 species, native of the Mediterranean Region, Himalayas, East Asia, the West Indies, and tropical Africa. Our collection is in the angle between the Lynch Walk and the Middle Walk.

B. sempervirens *L.*, Common Box. Native of South and West Europe. It is doubtfully native of the chalk in the south of England, and naturalised in other parts of Britain. The best natural or quasi-natural development on the chalk occurs in association with Yew on Box Hill, in Surrey. At Boxwell, in Gloucestershire, there is on the Oolite a fine boxwood, which is regularly cut and sold. The hard, heavy, close-grained wood, is very valuable for turning, carving, and wood-engraving. In Persian the word شمشاد is used for the Box, and sometimes for other plants.

B. balearica *Lamk*. Native of the Balearic Islands and South-west Spain. This is a much larger plant than *B.*

THE TREE OF HEAVEN (*AILANTHUS GLANDULOSA*),
NEAR THE STREAM (p. 76)

PLATE XX

XANTHOCERAS SORBIFOLIA, BESIDE THE
SUPERINTENDENT'S HOUSE (p. 83)

BUXACEAE

sempervirens, with larger less brightly-shining leaves. In its native land it sometimes reaches a height of 80 feet. It is much planted in gardens in the south of Europe.

Family CORIARIACEAE
CORIARIA

Coriaria, the only genus of this family, has 8 species, which are found in diverse parts of the world. The bed of Coriarias will be found on the north side of the South Walk between the openings of the Middle Walk and Broad Walk. There is a specimen of *C. myrtifolia* in Bay No. 5, and one of *C. terminalis* in Bay No. 6.

The floral formula of this genus is interesting. The members of the five floral whorls (K 5, C 5, A 5 + 5, G 5) alternate regularly, so that the carpels stand opposite to the *sepals*. This arrangement is called *diplostemony*. Compare carefully the flowers of *Coriaria* with those of *Geranium* or *Oxalis* in which there are also five whorls of floral leaves but the carpels are opposite to the *petals*. This is called *obdiplostemony*. In *Coriaria* the petals are keeled within, and eventually grow fleshy and enclose the carpels, forming a kind of false fruit.

C. myrtifolia *L.* This, the only European species, is native of the Mediterranean Region, where it is used in tanning and dyeing. The French call it *Carroyère*.

Family ANACARDIACEAE
COTINUS

Cotinus has two species, native of the North Temperate Zone.

C. coggygria *Scop.* (*Rhus cotinus* L.), Wig-tree, Venetian Sumach. Native of Middle and South Europe, and eastwards to China. The inflorescence branches copiously and many of its ramifications bear no flowers. As the fruits ripen the hairs on the inflorescence grow long and silky, and the manifold interlacing twigs produce the appearance

of a wig. This shrub yields the yellow dye-wood called
'young fustic,' which is especially suitable for dyeing leather.
Its twigs are used for tanning, and in the Himalaya for
basket-making, and various parts of the plant have been
used medicinally. The curious infructescence of the Wig-
tree was noticed by Theophrastus (*Hist. Plant.* III, xvi, 6)
and by Pliny (*N.H.* XIII, 41).

Rhus (Sumach)

Rhus has about 120 species, native of temperate and subtropical regions. The name is derived from the ῥοῦς of
Theophrastus (*Hist. Plant.* III, xviii, 5), which was probably
R. coriaria L., the only European species. Sumach is from
سُماق the Arabic name for the same species. Our collection
is on a bed between the stream and the Pond.

R. typhina *L.*, Staghorn Sumach, Vinegar Tree. Native
of Atlantic North America. This is the commonest *Rhus* in
cultivation in this country, where it is often seen growing in
small gardens. It is dioecious. The fruits are densely
packed in terminal infructescences, and covered with acid,
crimson hairs. The trivial name *typhina* suggests a resemblance of this infructescence to that of *Typha*; and one of
its popular names likens it to stags' horns. The other
popular name likens the flavour of the acid fruit to vinegar.
The bark, especially the root-bark, and the leaves, are rich
in tannin. Pipes for drawing the sap from the Sugar Maple
are made of the young shoots of this tree.

R. toxicodendron *L.*, Poison Ivy. Native of the United
States, in whose eastern parts it grows abundantly in hedgerows, thickets, and woods. The Poison Ivy is a shrub whose
leaves consist of three separate leaflets. It sometimes creeps
on the ground, and sometimes climbs up trees, rocks and
walls, to which it attaches itself by means of roots like the
climbing roots of the Ivy. It has no tendrils. Owing to its
acrid, milky juice, it is to many persons poisonous to the
touch, and sometimes causes very serious inflammation.
Occasionally indeed persons passing near the plants are

ANACARDIACEAE

poisoned without coming into contact with them. This is especially liable to occur at the time of flowering. The poisonous principle is insoluble in water, but is soluble in an alcoholic solution of lead acetate, with which affected parts should be washed as soon as possible.

Family CELASTRACEAE
EUONYMUS

Euonymus has 70 species, native of the North Temperate Regions. The Spindle Tree is the only British species. Each seed is partly or wholly encased in a fleshy, often beautifully coloured aril which is an outgrowth of the seedcoat. Our collection is on a long bed on the south side of the Border Walk, beyond the stream.

E. europaeus *L.*, Spindle Tree. Native of Europe (including Britain), West Siberia, and North Africa. In England it is common on chalk; it is rare in Scotland, and local in Ireland. The green, four-angled twigs and the pale crimson fruits set on slender stalks, and opening along four lines to expose the orange arils of the seeds, are striking features. The hard tough wood is excellent for turning, and makes the best charcoal for drawing. The seeds are poisonous.

E. verrucosus *Scop.*, Warty Spindle Tree. Native of East Europe to West Asia. It is commoner in some parts of the Continent than *E. europaeus*. The cylindrical branches of this species are densely warty. The smell of the flowers has been likened to that of *Geranium robertianum*.

E. atropurpureus *Jacq.*, Burning Bush, Wahoo. Native of the eastern and central United States. As our specimen shows, this species attains a greater size than most of the others. In Arkansas and eastern Texas it grows into a tree. Euonymus bark (*Euonymi cortex*) of the British Pharmacopoeia is the dried root-bark of this species.

E. latifolius *Miller*. Native of Europe. This species may be distinguished from the preceding ones by the parts of the flower being usually in fives instead of in fours. The beautiful rosy fruits have five wings.

CELASTRACEAE

E. japonicus *Thunb.*, Evergreen Spindle Tree. Native of Japan. This species has thick, dark, shining, evergreen leaves. It was introduced into this country in 1804. It is now often planted in gardens and is very commonly used for hedges at seaside places on the south coast of England.

Family ACERACEAE
ACER (MAPLE)

Acer has about 120 species, natives of the North Temperate Regions. Our collection of Maples extends from the west side of the West Walk into the north-west corner of the Garden. *A. ginnala* is on the bed between the West Walk and the Pond, and there are two fine specimens of *A. campestre* in the south-east corner of the Garden. The Maples have opposite, exstipulate leaves (see *Platanus*, p. 62, and *Liquidambar*, p. 60). Their winged fruits are well known. The leaves are often infested with honey-dew (see *Tilia*, p. 85). The sap of many Maples (especially *A. saccharum* Marshall, the Sugar Maple) abounds in sugar.

A. pseudoplatanus *L.*, Sycamore (and in Scotland 'Plane'). Native of Central Europe and West Asia. For many centuries this species has been thoroughly established in the British Isles, where self-sown trees abound. In gardens it often occupies space that might be devoted to more interesting trees. The leaves are generally spotted by the fungus *Rhytisma acerinum*. The timber is valuable for toys, shoe-makers' lasts, and the handles of brushes and brooms.

A. campestre *L.*, Common or Small-leaved Maple. Native of Europe and of North and West Asia. This is the only Maple indigenous in the British Isles, where it is common on good land and abundant on calcareous soils. It is not indigenous in Scotland. The Small-leaved Maple is usually a shrub and is very often coppiced, but it can, as our specimens show, grow into a handsome tree. Turners and cabinet-makers prize the roots and gnarled stems of this species. The leaves have five blunt lobes. The fruit-wings diverge horizontally in a straight line.

ACERACEAE

A. negundo *L.*, Box Elder. Native of North America, where it is very widely distributed. It is common in the forests that fringe the water-courses in the Prairie Region. This species has pinnate leaves and dioecious flowers. The variegated form, common in villa gardens, appeared as a sport. It is a carpellary plant. Our Box Elder is 50½ feet high, and would be a fine specimen if it had had room to develop. At its foot the ground is carpeted with Dwarf Bamboo (*Arundinaria vagans* Gamble).

A. monspessulanum *L.* Native of the Mediterranean Region. The small, neat leaves have three entire lobes. The slender leaf-stalks are as long as, or longer than the blades. In Persian it is called کی‌کف and کهوك.

A. creticum *L.* Native of the Eastern Mediterranean Region. This species differs from *A. monspessulanum* in being almost evergreen (in this country it loses its leaves at Christmas time), and in its leaf-stalks being much shorter than the blades, which tend to lose their two side-lobes and become narrow. The leaf appears to be intermediate between the broad deciduous Maple type and the narrow evergreen sclerophyll (see p. xiv).

A. platanoides *L.*, The Norway Maple. Native of Continental Europe from Norway southwards, and of the Caucasus. The leaves, whose lobes have almost parallel sides, and end in long points, are very handsome.

A. opalus *Miller.* Native of South Europe, the Caucasus, Persia, and Algeria. Like *A. campestre* it prefers calcareous soils. It flowers in March. The leaves of this species resemble those of the Sycamore, but the lobes are much shallower and blunter.

A. ginnala *Maxim.* Native of the Eastern Asiatic Region. The leaves have three lobes; the middle lobe is sharply serrate and much longer than the side ones. This tree shows very brilliant autumn colouring.

A. macrophyllum *Pursh.* Broad-leaved Maple. Native of eastern North America. It is most abundant and attains its greatest size in the humid climate of south-western Oregon.

ACERACEAE

Compare its huge leaves, which are sometimes a foot long, with those of *A. monspessulanum* and *A. creticum*, which live in dry climates.

Family HIPPOCASTANACEAE
AESCULUS

Aesculus has 16 species, native of the North Temperate Regions and of South America. Our collection is on the west side of the West Walk. Notice the large winter buds, which contain the young inflorescences.

A. hippocastanum *L.*, Horse Chestnut. Native of North Greece and Albania, the Horse Chestnut first reached Western Europe in 1576, when seeds were sent to Clusius in Vienna. Its timber is soft and of little value. The so-called 'nuts' of this tree are really seeds, and are at first enclosed in a fruit-wall whose surface is beset with prickles. Compare this with the nuts of the Sweet Chestnut which are real nuts (hard, one-seeded, indehiscent fruits), whose prickly case is equivalent to the 'cup' of the acorn.

In the section *Pavia* of the genus *Aesculus* the fruit has no prickles and is quite smooth. In the true *A. pavia* L. of North America, which is rarely seen in this country, both the tubular calyx, and the corolla, are bright red, and the stamens are not longer than the petals. The much commoner *A. carnea* Hayne, is probably a hybrid between *A. hippocastanum* and *A. pavia*. *A. octandra* Marshall (*A. flava* Ait.), of North America, has yellow flowers.

Family SAPINDACEAE

This family has more than a thousand species. Nearly all of them belong to the tropics, but the following two species, and several others, are natives of temperate regions.

KOELREUTERIA

Koelreuteria has 3 species, native of temperate China. Our specimen of *K. paniculata* is at the back of the Superintendent's House.

SAPINDACEAE

K. paniculata *Laxm*. Native of northern China. It has long been grown in Japan, and was introduced into this country in 1763. This tree has at least three charms: firstly, the graceful masses of foliage, which lead some to suppose that it is the tree represented on willow-pattern china; secondly, the wealth of deep yellow flowers produced in late summer; and thirdly, the curious bladder-like fruits marked by three furrows where the dissepiments draw the fruit-wall inwards. The seeds are made into rosaries. Lynch (p. 13) tells us that our specimen was planted in 1881 and that it was then about 4 feet high. In 1911 it was 29 feet high, in 1915 it was 30 feet, and it is now 31 feet high, and 37 feet 8 inches through.

XANTHOCERAS

X. sorbifolia *Bunge*, the only species, is native of northern China. Our specimen (Pls. XX and XXI) grows against the wall of the Superintendent's House, and is 30 feet high; an unusual height for this tree to attain in Britain, and perhaps in its native country too. It was 27 feet 3 inches high in 1915. The pinnate leaves with their daintily-cut leaflets are very beautiful. It flowers freely in May. The disc of the flower bears on its margin five yellow, horn-like projections. The seeds, which are nearly as large as hazel-nuts, are edible.

Family RHAMNACEAE

In this family the five stamens are usually hidden inside the five small, hooded petals, each like a little man in a tent.

PALIURUS

Paliurus has 2 species, native of South Europe and Asia. The fruit is winged, and contains a single stone. Specimens of Christ's Thorn will be found in Bay No. 5, and in the *Rhamnus* Collection.

P. australis *Gaertner*, Christ's Thorn. Native of South Europe and West Asia. It is the παλίουρος of Theophrastus, who gives it as an example of a typical shrub (θάμνος).

RHAMNACEAE

According to legends the crown of thorns was made either of this plant, or of *Zizyphus Spina-Christi*. The winged fruit resembles a broad-rimmed hat placed on a head, hence the French name 'Porte-chapeau.' The thorns are modified stipules; one of them is straight and the other recurved. Hedges are made of this plant in the south of Europe.

RHAMNUS (BUCKTHORN)

Rhamnus has about 100 species widely distributed. Two are natives of Britain. There is always more than one stone in the fruit. Our collection is on the west side of the Broad Walk.

R. catharticus *L.*, Common Buckthorn. Native of Europe, North Africa, and Siberia. This species in Great Britain prefers basic soils, and is a common component of the chalk scrub. The branches are thorny, and the leaves are narrowly ovate and sharply serrate. The flowers are dioecious.

R. frangula *L.*, Alder Buckthorn, Berry-bearing Alder. Native of Europe, North Africa, and Siberia. In Britain it is characteristic of woods on acid soils, but it grows together with *R. catharticus* on Wicken Fen. 'Frangula bark' was formerly in the British Pharmacopoeia. This species has no thorns. The leaves are obovate and entire, and the flowers are hermaphrodite.

R. purshianus *DC*. Native of western North America. 'Cascara sagrada' of the British Pharmacopoeia is the dried bark of this species collected at least one year before being used. This species differs from *R. frangula* in the leaves being downy, and the flower clusters being stalked.

R. alaternus *L.* Native of the Mediterranean Region. The leaf is a typical sclerophyll (see p. xiv) and resembles that of *Phillyrea*. (In *Phillyrea* the leaves are opposite; in *Rhamnus* they are usually alternate[1].) The flowers of this species have no petals.

CEANOTHUS

Ceanothus has about 36 species. They are all North American, especially Californian. Various species abound

[1] In *R. catharticus* the leaves are sometimes nearly or even quite opposite.

PLATE XXI

XANTHOCERAS SORBIFOLIA, A FLOWERING SPRAY (p. 83)

PLATE XXII

CLUSIA GRANDIFLORA, IN THE STOVE (p. 90)

RHAMNACEAE

in the sclerophyllous scrub known as 'chaparral.' Two common species of the chaparral are *C. cuneatus* Nuttall, which grows socially, and forms dense thickets of wide extent, and *C. thyrsiflorus* Eschs., the Californian Lilac. A specimen of the latter will be found in the Side Entrance. The fruit of *Ceanothus* is 3-lobed, and separates into three 2-valved nutlets.

Colletia

Colletia has 10 species native of South America.

C. cruciata *Gill et Hook.*, Anchor Plant. Native of South Brazil and Uruguay. The leaves are opposite, and in the axil of each there arise two buds, one above the other. The upper bud produces a large, triangular thorn, and the opposite pairs of these thorns resemble anchor-heads. The lower bud gives rise to a branch of unlimited growth and it is these branches that bear the flowers. In some species of the allied genus *Retanilla* both buds produce similar shoots. The Anchor Plant will be found on the Xerophyte Bed by the Lynch Walk.

Family ELAEOCARPACEAE

Aristotelia

Aristotelia has 7 species, native of the southern hemisphere.

A. macqui *L'Hérit.* Native of Chile. An evergreen shrub, which yields materials for a complete musical instrument. Its wood makes the body and its tough bark the strings. A medicinal wine is prepared from the berries. There is a specimen of this shrub amongst the Limes.

Family TILIACEAE

Tilia (Lime)

Tilia has about 10 species, native of the North Temperate Regions. The first leaf borne on the inflorescence of the Limes forms a pale wing. Honey-dew is the excretion of aphides which live on the leaves of Limes and other trees. Our collection of Limes is on the west side of the West Walk to the south of the Main Entrance.

T. cordata *Miller* (*T. parvifolia* Ehrh.), Small-leaved Lime. Native of Europe (not found in Greece or Turkey) and of Siberia. This is the only Lime that is certainly indigenous in the British Isles, where it is locally abundant in some ash woods. It is always a small tree. Except for minute beards in the axils of the nerves on the glaucous lower surface of the leaves, this species is quite hairless.

T. platyphyllos *Scop.*, Large-leaved Lime. Native of Europe. This species is probably not a native of Great Britain, though it occurs apparently spontaneously in some counties. It is closely allied to *T. cordata*, but has soft hairs on the twigs and on the lower surface of the much larger leaves. The fruit is prominently ribbed.

T. vulgaris *Hayne*, Common Lime. Very commonly planted. Its origin is not known. It is possibly a hybrid between *T. cordata*, which it resembles in being glabrous except for beards in the axils of the nerves beneath, and *T. platyphyllos*, which it approaches in its larger leaves and stature. The trunk of the Common Lime is often flanked by huge burrs, and its leaves are much infested by aphides, whose black honey-dew disfigures the pavement beneath its shade.

These three preceding Limes constitute the collective species *Tilia europaea* Linnaeus. The inner fibrous layer of the bark is the 'bast' of which mats, etc., can be made. An infusion of the flowers, which have a pleasant aroma, is an old-fashioned medicine still much used in France, and thus mentioned in White's *Natural History of Selborne*:—

'Dr Chandler tells, that in the south of France, an infusion of the blossoms of the lime-tree, *tilia*, is in much esteem as a remedy for coughs, hoarsenesses, fevers, etc., and that at Nismes, he saw an avenue of limes that was quite ravaged and torn in pieces by people greedily gathering the bloom, which they dried and kept for these purposes. Upon the strength of this information we made some tea of lime-

TILIACEAE

blossoms, and found it a very soft, well-flavoured, pleasant, saccharine julep, in taste much resembling the juice of liquorice.'

T. tomentosa *Moench* (*T. argentea* Desf.), White Lime. Native of South-east Europe. The mature leaves of this species are dark green above and covered with silvery felt beneath. The flowers smell of honey, but have not the characteristic Lime aroma.

T. petiolaris *DC.* Origin unknown. It differs from *T. tomentosa* in having longer leaf-stalks, and in the leaf-blades being more finely and regularly serrate, as well as in the fruit. Of this tree we have a magnificent specimen, whose lower branches sweep the ground. Before it puts out its leaves in the spring, the ground beneath it is carpeted by a mass of a yellow-flowered South European Umbellifer, *Smyrnium perfoliatum* L.

T. americana *L.*, Basswood. Native of eastern and central North America. The leaves are very large (sometimes more than a foot in length) and quite hairless. The stipules are large and broadly ovate.

Family MALVACEAE

HIBISCUS

Hibiscus has about 150 species, natives of the tropical and subtropical regions of both hemispheres. *H. rosa-sinensis* and other species are cultivated in tropical gardens. There is a bed of *H. syriacus* on the south side of the Lynch Walk.

H. syriacus *L.* Native of China and India. It is the *Althaea frutex* Hort. (Compare the epicalyx and fruit of *Hibiscus* and *Althaea.*) It has been grown in this country for more than three hundred years. Both Linnaeus's trivial name and the earlier popular name 'Syrian Ketmie' suggest that it came to us from Syria, where it is cultivated. 'Ketmie' is doubtless from Arabic خطمى, a name used for the Marsh-mallow (*Althaea officinalis* L.) wherever Islamic culture has spread.

MALVACEAE

H. trionum *L.* Native land uncertain. It is common in many warm regions. This is an annual with splendid purple-eyed flowers which open only for a few hours, whence Gerard called it 'Malva horaria.'

Family STERCULIACEAE

In this family the outer whorl of stamens is either absent altogether or represented by staminodes.

FREMONTIA

There is only one species of this genus. Our specimen grows at the west entrance to the Houses.

F. californica *Torrey*, Slippery Elm. Native of the southern slopes of the Californian mountains. This plant is one of the few extra-tropical *Sterculiaceae*. The brightly-coloured calyx forms the conspicuous part of the flower; the corolla is absent. In its native land this plant attains a height of from 20 to 30 feet with a trunk a foot or more in diameter. The mucilaginous inner bark is sometimes used for poultices. Note the stellate hairs on the leaves.

THEOBROMA

Theobroma has 20 species, native of tropical America. Our specimen is in the Stove.

T. cacao *L.*, Cacao. The seeds of this tree yield cocoa, chocolate, and cacao butter. Chocolate was the great drink of the Aztecs and Incas. Montezuma, Emperor of the Aztecs, "took no other beverage than the chocolatl, a potation of chocolate, flavoured with vanilla and other spices, and so prepared as to be reduced to a froth of the consistency of honey, which gradually dissolved in the mouth. This beverage, if so it could be called, was served in golden goblets, with spoons of the same metal or tortoise-shell finely wrought. The Emperor was exceedingly fond of it, to judge from the quantity—no less than fifty jars or pitchers being prepared for his own daily consumption! two thousand more were allowed for that of his household[1]."

[1] Prescott's *Conquest of Mexico*.

STERCULIACEAE

The flowers and fruits are borne on the trunk and main branches. The five fertile stamens are enclosed in the curiously-hooded petals. The five staminodes stand up in the midst of the flower.

For a very interesting and beautifully-illustrated account of this plant see *Cocoa and Chocolate, Their History from Plantation to Consumer*, by A. W. Knapp.

Family CAMELLIACEAE

CAMELLIA

Camellia has 16 species, native of India, China, and Japan. Those members of the genus which have large sessile flowers whose calyx consists of numerous sepals, are always called *Camellia*, whereas the tea plant (*q.v.*) is sometimes put into a distinct genus *Thea*. A fixed oil is expressed from the seeds of several of the species, including the tea plant. Our specimens will be found in the Corridor, and in Bay No. 3.

C. japonica *L.*, Common Camellia. Native of China and Japan. The leaves are very smooth and shining, and the flowers are solitary at the ends of the branchlets. This species was introduced in the eighteenth century. It has long been a favourite garden plant. All the better known varieties are said to have arisen spontaneously about the year 1792. The flowers of some of them look like turnip-carvings.

C. reticulata *Lindl.* Native of China, introduced in 1820. This species is peculiar in having a dull leaf surface. It is by far the most magnificent of the Camellias. The flowers, which are in groups of twos and threes, are more than six inches across, and look like Paeonies.

C. sasanqua *Thunb.* Native of China and Japan. The leaves are smaller and narrower than those of the first two species, and their stalks and midribs are hairy. The leaf-blades are dark and shining. The flowers do not exceed two inches across.

CAMELLIACEAE

C. theifera, Tea Plant[1]. Native of Upper Assam. It has been cultivated for many centuries in China. In the year 1851 Chinese cultivated varieties were introduced into India, where tea cultivation has extended widely. It is grown also in Ceylon and Japan. The Tea Plant has nodding white flowers borne on stalks. The sepals are five in number and persistent. To make black tea the leaves are fermented before being dried. Green tea is dried without fermentation. The branches of the midrib of the leaf meet to form a vein which runs parallel with the leaf margin. This marginal vein, which can be seen in leaves taken from the tea-pot, distinguishes tea-leaves from the leaves of willows and other plants with which tea used to be adulterated.

Family GUTTIFERAE

Guttiferae means drop-bearers, and the family owes its name to the thick, yellow or greenish gum-resin that drips slowly from the cut stems or bark of many of its members. Gamboge is the gum-resin of species of *Garcinia*, used as a medicine and pigment. The British genus *Hypericum* (St John's Wort) belongs to the *Guttiferae*.

CLUSIA

Clusia has about 100 species, native of tropical America. They are trees or shrubs often growing upon other plants, which they strangle with their roots. Our specimens are in the Stove. There is a portrait of the great botanist Clusius (1525–1609), after whom the genus is named, on the right as you enter the right-hand door of the Botany School.

C. grandiflora *Splitg.* (Pl. XXII). Native of Guiana. The flowering of this plant, which stands in the centre of the

---

[1] The common hedge plant and cottage ornament, *Lycium chinense* Miller (Family *Solanaceae*), is often called the Chinese Tea Plant. It was sent from China along with the true Tea Plant with which it was subsequently confused.

GUTTIFERAE

Stove, is an occurrence that excites the wonder even of the most incurious. For weeks beforehand the large swollen buds look ready to open. Each flower blooms for a very short time, but during that time the large roseate diaphanous petals show a splendour, which, if equalled, is certainly not surpassed by any other flower. Our specimen is staminate. The united filaments in this species form a shallow cup, in the centre of which is a mass of staminodes embedded in gum-resin. Note the tangle of roots about the base of the stem.

C. flava *L*. Native of Jamaica, where it grows abundantly on rocks. A yellow gum-resin flows from every part of the plant when cut. The Caribs used this, mixed with tallow, for caulking their boats, and it is sometimes used as a dressing for sores in horses.

Family CISTACEAE

Cistus (Gum Cistus)

The 20 species of Cistus grow in the Mediterranean Region and in the Canary Islands. Unlike the typical Mediterranean woody vegetation they are not sclerophyllous plants. They differ from *Helianthemum* in their five-, or sometimes ten-valved capsules, and in never having yellow petals. *Cistus creticus* and other species yield the deliciously fragrant exudate called Gum Ladanum. It is collected by dragging through the shrubs, in the heat of the day, a kind of rake with long, leathern prongs, and it is still, as in old times, gathered from the beards of goats which browse on the foliage (see Pliny, *N.H.* xxvi, 30). The Greeks during the great war used ladanum as a dressing for wounds. Our collection occupies a bed on the north side of South Walk.

Family FLACOURTIACEAE

The members of the family *Flacourtiaceae* are nearly all tropical, but we have specimens of three species which are hardy in this country.

FLACOURTIACEAE

Azara microphylla *Hk.f.* Native of Chile. This tree appears to have pairs of leaves of which one is much larger than the other. In reality the leaves are alternate, and the smaller leaf of each pair is a single stipule. A specimen will be found at the east end of the Border.

Idesia polycarpa *Maxim.* Native of China and Japan. In its native land it is a large tree, whose straight trunk is crowned with numerous horizontal branches. There is a specimen on a bed between the West Walk and the Pond.

Carrierea calycina *Franchet.* Native of West and Central China. It was introduced into this country in 1908. In its native land it sometimes grows into a tree 40 feet high and is said to look very beautiful when bedecked with its clear bluish-white flowers. We have a specimen near *Idesia polycarpa*.

Family CACTACEAE

The 1500 species of this family are, with very few exceptions, natives of the New World. These exceptions are some species of *Rhipsalis*. *R. cassytha* Gaertn. is widely distributed in both tropical America and Africa, and is found also in Ceylon. Nearly all our *Cactaceae* are in the Second Succulent House. They should be studied with the help of the excellent account of the family in Willis.

Opuntia

The 150 species of Opuntia are all natives of the New World. The number of species and of individuals is greatest in the south-west United States and in North Mexico, where they form the most conspicuous part of the flora. Some are found as far north as Canada. Wherever they have been introduced into the Old World they have become troublesome weeds. Some species (especially *O. ficus-indica* Mill., and *O. tuna* Mill.) are cultivated for their edible fruits which are called 'prickly pears.' The finest fruits are produced in Sicily. The small, fleshy leaves of the Opuntias usually fall very early. The thick, often flat, joints of the

PLATE XXIII

OPUNTIAS IN THE OPEN BY THE LYNCH WALK (p. 92)

CACTACEAE

stem are beset with tufts of bristles and barbed spines. They are good hedge plants and spineless forms make valuable fodder. Most of our Opuntias are in Succulent House No. 2. There is also a collection by the Lynch Walk (Pl. XXIII), which is protected during winter.

O. cantabrigiensis *Lynch*. The origin of this species is unknown. Lynch, who first described it (1903), said that for many years the plant had been a feature in this Garden.

Family MYRTACEAE

Myrtus

Myrtus has about 70 species. All the species except the Myrtle are natives of tropical and subtropical regions. Our specimens of the Myrtle are in Bay No. 4.

M. communis *L.*, Myrtle. The Common Myrtle occurs in the Mediterranean *garigue* and has been well known in the south of Europe from classical times ('amantes litora myrtos,' Virg.). It has been asserted without evidence to have been introduced from West Asia. The Myrtle was held sacred to Venus (Hor. *Odes*, 1, 38, 5), and wreaths of it were worn by Athenian magistrates and by victors in the Olympic games. The pellucid dots on the leaves are intercellular spaces containing the volatile oil which gives the plant its sweet scent. In Persian, the Myrtle is called مورد, a word probably borrowed from the Greek μύρτος.

Pimenta

Pimenta has 5 species, native of warm regions. Our specimen of the Allspice is in the Stove.

P. officinalis *Lindl.*, Allspice, Pimento. Native of the West Indies, Mexico, and South America. The dried unripe berries are the Allspice of commerce, so called because their aroma seems to combine the fragrance of cinnamon, cloves, and nutmeg. The chief supplies come from Jamaica, where the trees are planted in groves called Pimento Walks. Both Allspice (*Pimenta*) and its volatile oil (*Oleum pimentae*) were in former British Pharmacopoeias, but neither is now official. The oil is a constituent of Bay Rum.

CORNACEAE

Family CORNACEAE

CORNUS (CORNEL, DOGWOOD)

Cornus has about 40 species, native of the northern hemisphere and the Andes. There are two species in Britain, *C. sanguinea* L. (Dogwood), and *C. suecica* L. (Dwarf Cornel). The two sorts of Cornel described by ancient writers and early herbalists are 'Cornus mas,' which has retained its name, and 'Cornus femina' which is our *C. sanguinea*.

C. mas *L.*, Cornelian Cherry. Native of Europe and North Asia. This species in February, before the leaves appear, is covered with masses of bright yellow flowers, which are arranged in small umbels. Each umbel is subtended by four greenish bracts which resemble a calyx, but are not, of course, in the position of a calyx. The fruits look like cherries, but, because they are borne by a Cornel, and not by a *Prunus*, are called 'Cornelian Cherries.' On the Continent they make jam of them and sometimes eat them raw. Compare the different parts of this fruit with those of a real cherry.

C. florida *L.*, Flowering Dogwood (Lat. *floridus*, blooming, flowering). Native of eastern North America, where it is very common in the forests of the middle and southern States. It sometimes reaches a height of 40 feet, and likes to grow beneath the shade of taller trees. The so-called 'flowers' of this species are really inflorescences similar to those of the Cornelian Cherry. The real flowers are densely clustered and inconspicuous, but the four bracts beneath them are white and petal-like, and each of them measures up to two inches in length.

C. suecica *L.*, Dwarf Cornel. Native of Northern and Arctic Europe, Asia, and America. In Great Britain it is found in acid soils from Yorkshire to Sutherland. This species is a lowly herb. The four bracts are white and about a quarter of an inch long.

C. sanguinea *L.*, Dogwood (formerly Dogberry-tree, dog

CORNACEAE

meaning unfit for human food). Native of Europe, North and West Asia, and the Himalayas. In England it is abundant on the chalk, but is not uncommon on other soils. It does not extend north of Westmorland. In autumn the stems and leaves are dark red. The young branches make excellent skewers. The flowers are arranged in bractless cymes.

C. alba *L.* Native of Siberia and China. The bark of the young shoots of this species reddens in autumn, and through the winter keeps up a warm glow. The bed of flaming stems among the Cornels is maintained by coppicing the plants to make them produce as much new growth as possible. The flowers of this species are arranged, like those of our Dogwood (*C. sanguinea* L.), in cymes with no bracts at the base.

AUCUBA

Aucuba has 3 species, native of China, Japan, and the Himalayas.

A. japonica *Thunb.*, the so-called Japanese Laurel, is a dioecious shrub with leaves of the laurel type[1]. A variety with variegated leaves is much cultivated.

HELWINGIA

Helwingia has 2 species, 1 native of the mountains of India, the other one of Japan. Our specimen of *Helwingia rusciflora* will be found in Bay No. 2.

H. rusciflora *Willd.* Native of Japan. Note the alternate leaves. The leaves in *Cornaceae* are usually opposite. The axillary peduncle is adnate for its entire length to the stalk and midrib of the leaf, so that the flowers and fruits spring apparently from the middle of the leaf-blade.

---

[1] The leaves are opposite. In *Laurus* they are alternate.

ERICACEAE

METACHLAMYDEAE

Family ERICACEAE

DABEOCIA

Dabeocia has only 1 species. Specimens of it will be found in the bed of plants which grow on acid peat. This bed lies immediately to the west of the Rock Garden.

D. polifolia *Don*, St Dabeoc's Heath. This is a member of the British flora whose geographical distribution is very interesting. It grows abundantly in the south-west of Ireland, and nowhere else in the British Isles. Outside the British Isles it is found in the west of France, in Spain, in Portugal, and in the Azores. Members of our flora with this 'Atlantic distribution' are found in the west of Ireland and in the south-west of England.

ARBUTUS (STRAWBERRY TREE)

Arbutus has 20 species, native of the warmer parts of the northern hemisphere.

A. unedo *L.*, Common Strawberry Tree. Native of the Mediterranean Region and of Ireland, where it grows abundantly and very luxuriantly in woods of *Quercus sessiliflora* Salisb., in Co. Kerry. Apart from an isolated thicket on rocky cliffs by the Trieux River in Côtes-du-Nord, a distance of 600 miles separates the Irish Strawberry Trees from their nearest sisters in western France. This species and *A. andrachne* L. (which has entire leaves) are characteristic trees of the Mediterranean sclerophyllous woodland called *macchia*. Their leaves are intermediate between the sclerophyll and the laurel type.

A. menziesii *Pursh.*, Madrona Laurel: its leaves are of the laurel type. This species is a component of the Californian sclerophyllous woodland, and just as the common Strawberry Tree extends out of the Mediterranean sclerophyllous zone up Western Europe, the Madrona Laurel creeps up the Pacific coast of America as far north as British Columbia. This is one of several interesting analogies between Mediterranean and Californian vegetation.

ERICACEAE

This tree must be the largest member of its family. It reaches a height of 80–100 feet with a tall, straight trunk 4–7 feet in diameter. The branchlets after their first winter assume a bright reddish-brown colour.

Family EPACRIDACEAE

This family consists of about 340 species of sclerophyllous shrubs, all of which are native of the southern hemisphere. Most of them grow in temperate Australia and Tasmania, where they seem to take the place of the heaths of South Africa, to which they are allied.

STYPHELIA

Styphelia has 170 species, native of Australia, New Zealand, Sandwich Islands, New Caledonia, and India.

S. richei *Labill.* (*Leucopogon richei* R. Br.). This species, of which we have a good specimen in the Temperate House, grows very abundantly in the temperate parts of Australia. Its narrow, hard, greyish, sharp-pointed leaves finely lined by delicate veins, are typical of the family to which it belongs. In the *Botanical Magazine* (T. 3251), there is an interesting account of the painful expedition near Cape Le Grand, South-west Australia in 1792, on which this plant was discovered. The curious leguminous shrub *Chorizema ilicifolium* Labill. (also in the Temperate House) was discovered on the same expedition.

Family EBENACEAE

DIOSPYROS

Diospyros has about 200 species, most of which are confined to the tropics. Ebony is the hard, black heart-wood of several tropical species. Our specimen of *D. kaki* is in the Corridor, and our best specimen of the Persimmon is on the east side of the West Walk.

D. virginiana *L.*, Persimmon Tree. Native of eastern North America, extending as far north as Connecticut. It is very abundant in the South Atlantic and Gulf States, where, spreading by means of its creeping roots, it covers abandoned fields with shrubby growth. In the primeval forest it sometimes reaches a height of 115 feet. The fruit is eaten in great quantities in the Southern States. It is very astringent even when ripe, but becomes sweet and edible when softened by frost. The hard wood is valuable for turning, and the bitter, astringent bark has been used medicinally.

D. kaki *L.f.*, Chinese Persimmon, Kaki. Native of China and perhaps also of Japan. This species is now cultivated in many countries for its fruit, which is delicious when dead ripe.

Family OLEACEAE

The flowers of the Oleaceae have only two stamens.

FRAXINUS (ASH)

Fraxinus has about 40 species, native of the northern hemisphere. The leaves are opposite and usually pinnate. The winged fruits are often called 'keys' (see *Ailanthus*, p. 75). Our collection is on the south side of the South Walk.

F. excelsior *L.*, Common Ash. Native of Europe and North Africa. In England ashwoods are characteristic of the limestone hills of the North and West (see *Fagus*, p. 43). Typical ash woods occur in Yorkshire, Derbyshire, Westmorland and Somerset.

"It is probable that at some time, the whole of the limestone areas of England below about 1000 feet (*c.* 300 m.), or perhaps even 1250 feet (*c.* 380 m.), were covered by a primaeval ash forest, just as similar places on the older siliceous hills were once covered by forests of oak (*Quercus sessiliflora*) and birch (*Betula tomentosa*). The numerous place-names including the word 'ash' indicate that the abundance of *Fraxinus excelsior* is of long standing. In north Derbyshire, for example, there are Ashwood dale, Ashford, Money Ash (= 'many ash'), and, on the edge of the plateau at the upper limit of woodland, One Ash[1]."

[1] Moss in Tansley, p. 148.

OLEACEAE

The Ash is also a frequent, and sometimes an important component of other types of wood. It grows in all types of soil except poor sands and peat, and is especially frequent in wet places and by stream sides. After Hazel the Ash is the commonest component of our coppiced woods.

The Teutonic peoples have innumerable superstitions connected with this tree; the famous Ygdrasil of Scandinavian mythology was an ash. See *Fraxinus* in *Treasury of Botany*.

The flowers in this species have neither sepals nor petals. The tough timber is excellent for many purposes.

F. ornus *L.*, Manna Ash, Flowering Ash. Native of the Mediterranean Region. The flowers of this species have both sepals and petals. The dried sugary exudation from the stem is the 'Manna' formerly in the British Pharmacopoeia, but not the Manna of the Bible.

PHILLYREA

Phillyrea has about 6 species, native of the Mediterranean Region. Our collection forms a beautiful, evergreen thicket on the west side of the Broad Walk.

P. angustifolia *L.*, **P. media** *L.*, and **P. latifolia** *L.* These three species of *Phillyrea* occur in great abundance in the Mediterranean Region. The leaf is a typical sclerophyll (see p. xiv).

P. decora *Boiss.* Native of Lazistan. This species has long, narrow leaves very unlike those of the western species.

OLEA

Olea has 35 species, native of the Mediterranean Region, Africa, and New Zealand. Our Olive Trees are in Bay No. 1 and the Temperate House.

O. europaea *L.*, Olive Tree. From early ages the Olive has been cultivated in the Mediterranean Region, and was intimately connected with early civilisation. It is the זַיִת of the Old Testament, where it figures as the symbol of reconciliation, and of peace and plenty, and the زيتون of Islamic literature. Theophrastus calls it ἐλαία and gives it as an example of a tree (δένδρον). Compare *Paliurus*, p. 83. The wild olive is thorny and its fruit is worthless. It is

OLEACEAE

used as a stock upon which is grafted the cultivated form (var. *sativa* DC.), from the flesh of whose drupe the well-known oil is expressed. St Paul speaks of the grafting of the Olive in Rom. xi, 17–24.

The עֵץ שֶׁמֶן (oil-tree) of the Old Testament, which is translated 'Pine branches,' 'Olive Tree,' and 'Oil Tree,' is undoubtedly the Oleaster (*Elaeagnus angustifolia* L. Fam. *Thymelaeaceae*). See Tristram, p. 371.

Family APOCYNACEAE

NERIUM (OLEANDER)

Nerium has 2 or 3 species, whose range extends from the Mediterranean Region to East Asia. The leaves of the Oleanders are usually in whorls of 3. Our specimens of the Common Oleander are in the Temperate House and the Corridor.

N. oleander *L.*, Common Oleander. Native of the Mediterranean Region. For an account of its abundance and beauty in Palestine see Tristram, p. 416. The Indian Oleander, كنير, has fragrant flowers, and is usually considered to be a distinct species (*N. odorum* Soland). Its Sanskrit name अश्वमारक means 'horse-killer,' and the Arabic سمّ الحمار, Persian خرزهره, and Italian *Ammaza l'asino*, names for the Common Oleander, all mean 'ass-poison.' The plant is very poisonous and all its parts are used medicinally in the East. Jealous Indian women destroy themselves by eating the roots.

Family ASCLEPIADACEAE

MARSDENIA

Marsdenia has about 70 species. All of them except *M. erecta* are natives of the tropics. Our specimen of *M. erecta* will be found in the Side Entrance.

M. erecta *R. Brown*. Native of South-east Europe and Asia Minor. The latex is very poisonous and will blister the skin. Formerly the leaves were employed as a medicine under the name *Herba Apocyni folio subrotundo*.

Family VERBENACEAE
VITEX

Vitex has about 100 species. Nearly all of them are tropical.

V. agnus-castus *L.*, Chaste Tree, Abraham's Balm. Native of the Mediterranean Region; cultivated in Britain in 1570. Throughout the ages this plant has been associated with chastity. It is the ἄγνος of the Greeks and the *Vitex* of the Romans (see Pliny, *N.H.* XXIV, 38). The women of ancient Greece strewed their beds with it during the *Thesmophoria*, a festival during which chastity was enjoined. In Islamic medicine it is called اثلق, and پنج انگشت ('five fingers'). The dried berries under the names of حب الفقد, and سنبهالوکے بیج (cf. Hindi संभालना, to restrain) are imported into India for medicinal purposes. In the East the plant is also called ذو خمسة اوراق ('the five-leaved') and كف مريم ('the hand of Mary').

Family SOLANACEAE
ATROPA

Atropa has 2 species, native of Europe, the Mediterranean Region, and Asia. The Deadly Nightshade will be found on the Medicinal Bed. It also grows wild under the shade of trees in several parts of the Garden.

A. belladonna *L.*, Deadly Nightshade. Native of Europe, Persia, and the Western Himalaya. In Britain it is probably native in the edges of woods and scrub on the southern chalk. It was formerly much cultivated, especially near Cambridge, and many of the wild plants now found are escaped relics of cultivation. All its parts are poisonous, and the attractive, cherry-like fruits are sometimes eaten with fatal results. Though long known as a poisonous plant, its use in medicine is of comparatively recent date. The name *belladonna*, first used in botanical books by Clusius in the sixteenth century, means in Italian 'fair lady,' and was

SOLANACEAE

applied, perhaps because Italian ladies used to dilate the pupils of their eyes with it, or perhaps because Leucota, a famous poisoner, used Deadly Nightshade for destroying his fair victims. The plant was first called *Atropa* (Ἄτροπος, one of the Fates) by Linnaeus.

Our native species of *Solanum*, which have small flowers like those of the potato, are often mistaken for Deadly Nightshade. The corolla of the Deadly Nightshade is a greenish-purple bell an inch in length. The berries of the Deadly Nightshade, each standing in its conspicuous, spreading calyx, resemble the fruit of no other British plant.

MANDRAGORA

Mandragora has 3 species, native of the Mediterranean Region and the Himalaya. Our specimen of *M. officinarum* will be found in Bay No. 2.

M. officinarum *L.*, Mandrake. Native of the Mediterranean Region. The mandrake was formerly an object of much superstition. Since death was said to overtake anyone who pulled up its roots, which strongly resemble the human form, they used to tie a hungry dog to the plant and loosen the earth around it. The dog, struggling to obtain a piece of meat, which was held slightly out of his reach, tore up the roots, but was said to perish immediately afterwards. The plant was believed to shriek when torn from the ground. Mandragora contains a mydriatic alkaloid, and was perhaps used as an anaesthetic. Another method of gathering the plant is mentioned by Theophrastus (*Hist. Plant.* IX, viii, 8). In Persian the mandrake is سگ‌کن, مهرگیاه ('dog-dig'), and مردم‌گیاه ('man-plant'), in Arabic يبروح, and in Hebrew דּוּדָי. See articles in dictionaries, and *The Treasury of Botany*, II, 716. In America a totally different plant *Podophyllum peltatum* L. (Family *Berberidaceae*) is often called Mandrake.

SCROPHULARIACEAE

Family SCROPHULARIACEAE

VERONICA (SPEEDWELL)

Veronica has about 200 species, native of the North Temperate Regions, and of New Zealand and Australia. The 17 British species, which are all herbaceous, are called Speedwells.

Veronica is the largest genus of flowering plants in New Zealand, where there are about 84 species, of which all but 3 are endemic. A collection of New Zealand Veronicas will be found in Bay No. 1.

V. cupressoides *Hk.f.* Native of New Zealand. The small, scale-like leaves give the twigs of this plant a likeness to those of a Cypress. The nine species of *Veronica* which have this form are sometimes called Whip-cord Veronicas. They are all endemic in New Zealand. The seedlings of these plants bear spreading, more or less lobulate, leaves, and adult plants sometimes produce leaves of this kind.

Family RUBIACEAE

COFFEA

Coffea has 40 species, native of the tropics of the Old World. Our Coffee Plant is in the Stove.

C. arabica *L.*, Coffee Plant. Probably indigenous only in certain hilly regions of Abyssinia, the Soudan, Guinea, and Mozambique. Eastern names of coffee can all be traced from either قهوة, which originally meant wine prepared from the pulp of the fruit ('Coffee' is derived from this word), or بُن the Abyssinian name of the plant. The early history of coffee is obscure. It came into Europe during the seventeenth century. Up to 1690 the world's supply came from Arabia and Abyssinia. To-day the chief supplies come from Brazil, where it is extensively cultivated.

The fruit is a drupe containing two stones. The seeds are the well-known coffee 'beans.' Notice the interpetiolar stipules characteristic of the family *Rubiaceae*. In the

RUBIACEAE

British *Rubiaceae* (tribe *Galieae*) the stipules are as large as the leaves, and leaves and stipules together form star-like whorls.

MYRMECODIA

Myrmecodia has 20 species, native of the Indo-Malay Region. They are all epiphytes, *i.e.* they grow free from the ground attached to the stems of other plants from which, however, they obtain no food. At the base of the stem is a large corky tuber whose interior consists of a labyrinth of intercommunicating galleries inhabited by ants. The ants are not known to be of any use to the plants, cf. *Acacia spadicigera*, p. 66. Our specimen of *M. echinata* is in the Orchid House.

Family CUCURBITACEAE

Most of the members of this family are annual herbs climbing by tendrils. They grow remarkably fast. The Gourds on the herbaceous beds are trained over trellises, and in August they produce their curious fruits. Some tropical species with beautifully coloured fruits grow in the Aquarium.

Ecballium elaterium *A. Rich.*, Squirting Cucumber, the only species of the genus, is native of the Mediterranean Region, where it affects deserts and waste places. It has no tendrils. The wall of the ripe fruit is tense and the seeds are contained in a mucilaginous fluid under pressure. When the fruit falls a hole is left where the stalk was attached to it, and the sudden contraction of the tense fruit-wall squirts the fluid containing the seeds through this hole with sufficient force to deposit the slimy mass several yards away from the plant. *Elaterium*, which was formerly in the British Pharmacopoeia, was the dried sediment of the juice of the fruit. Its active principle called *elaterin*, a violent hydrogogue cathartic, was given in doses of 2–6 milligrams ($\frac{1}{40}-\frac{1}{60}$ grain). In the East the unripe sliced fruits are used for malaria and other diseases. A former name for the plant was *Cucumis asininus*, and the Persian خرخیار and

CUCURBITACEAE

Arabic قثاءالحمار both mean 'Ass's Cucumber.' Other Persian names are کموزکـوَل and کموزسگی. This plant will be found on the Medicinal Bed.

Family COMPOSITAE
INULA

Inula has 90 species, native of Europe, Asia, and Africa. Elecampane will be found in the Side Entrance and on the Medicinal Bed.

I. helenium *L.*, Elecampane. Native probably from Central Europe to Persia. It has been cultivated in Europe, including Britain, from the Middle Ages for its rootstock, which was formerly a well-known household remedy for coughs and colds.

> Excellent herbs had our fathers of old—
> Excellent herbs to ease their pain—
> Alexanders and Marigold,
> Eyebright, Orris and Elecampane.   (KIPLING.)

Elecampane, which is still sometimes eaten candied, contains a bitter principle called *helenin* said to be destructive to the tubercle bacillus. The white, velvety hairs on the lower surface of the leaves distinguish this plant from *Buphthalmum speciosum* Schreb., with which it is sometimes confused. Elecampane has escaped from cultivation and is quite naturalised in Britain, North Europe, and North America. The carbohydrate *inulin* was first obtained from this plant.

CHRYSANTHEMUM

Chrysanthemum has about 200 species, native of the northern hemisphere. The garden 'Chrysanthemums' have been derived from *C. morifolium* Ram. (*C. sinense* Sabine), and *C. indicum* L. Costmary grows in the Side Entrance.

C. balsamita *L.*, Costmary, Alecost. Native of Asia Minor. It has long been cultivated as a herb, and was formerly added to ale and negus. It is said to have been introduced into this country, from Italy, in 1568. The leaves, like those of other Chrysanthemums, are aromatic, but their fragrance is like that of a mint rather than of a Chrysanthemum. The variety generally cultivated has no ray florets, but there is a variety in Armenia and North Persia with showy, white rays. In the Middle Ages the plant was associated with the Virgin Mary, and the word Costmary is derived from O.E. 'cost,' a fragrant root (Greek κόστος, Arabic قُسْط, Sanskrit कुष्ठ), and 'Mary'; and not, as is sometimes stated, from *costus amarus* (bitter cost).

INDEX

Abele, 32
Abraham's Balm, 101
Abyssinian Banana, 30
Acacia, 3, 65, 66
　armata, 66
　dealbata, 66
　decurrens, 66
　pycnantha, 66
　spadicigera, 66
Acer, 80
　campestre, 80
　creticum, 81, 82
　ginnala, 80, 81
　macrophyllum, 56, 81
　monspessulanum, 81, 82
　negundo, 81
　opalus, 81
　platanoides, 81
　pseudoplatanus, 50, 80
　saccharum, 80
Aceraceae, 80
Acorus, 20
　calamus, 20
Adam's Needle, 27
Aegilops, 44
Aegle, 75
　sepiaria, 75
Aesculus, 82
　carnea, 82
　flava, 82
　hippocastanum, 82
　octandra, 82
　pavia, 82
Agave, 28
　parryi, 29
　salmiana, 29
　utahensis, 29
　vera-cruz, 29
Ailanthus, 75
　glandulosa, 76
　moluccana, 76

Ailanthus—*contd.*
　vilmoriniana, 76
Alder, 42
　Buckthorn, 84
　Common, 42
Aldrovanda vesiculosa, 59
Alecost, 105
Aleppo Pine, 6
Allspice, 93
Almond, 64
Almond-leaved Willow, 36
Alnus, 42
　cordifolia, 42
　glutinosa, 42
　incana, 42
Alocasia, 23
Aloe, 26
　Common, 26
Aloë, 26
　barbadensis, 26
　chinensis, 26
　ciliaris, 27
　perryi, 26, 27
　plicatilis, 27
　vera, 26
Althaea, 87
　frutex, 87
　officinalis, 87
Amaryllidaceae, 28
American Aloe, 28
Anacardiaceae, 77
Ananas, 25
　sativus, 25
Anchor Plant, 85
Andropogon, 15
　halepensis, 15
　nardus, 15
　sorghum, 15
Anonaceae, 5, 55
Apocynaceae, 51, 100
Apple, 63

INDEX

Apricot, 63, 64
Aquillaria agallocha, 26
Araceae, 19, 23
Arbor-vitae, 9
Arbutus, 96
  andrachne, 96
  menziesii, 96
  unedo, 96
Aristotelia, 85
  macqui, 85
Arrow Arum, 22
Arundinaria vagans, 81
Asclepiadaceae, 100
Ash, 98, 99
  Common, 98
Asimina, 55
  triloba, 55
Aspen, 32, 33
Astragalus, 71, 72
  gummifer, 72
  tragacantha, 72
Atlantic Cedar, 5
Atropa, 101
  belladonna, 101
Aucuba, 95
  japonica, 95
Austrian Pine, 7
Azara microphylla, 92

Banana, 30
Banyan, 50
Barbados Aloe, 26
Barberry, Common, 52
Basswood, 87
Bauhinia, 67, 68
  variegata, 68
  yunnanensis, 67
Bay Laurel, 56
Beech, 43
Berberidaceae, 51, 102
Berberis, 51
  aquifolium, 52
  darwinii, 52
  empetrifolia, 52
  fortunei, 52
  fremontii, 52
  stenophylla, 52

Berberis—*contd.*
  vulgaris, 52
Berry-bearing Alder, 84
Betula, 40
  alba, 41
  nana, 40, 41
  papyrifera, 41
  pubescens, 41
  tomentosa, 40, 98
  verrucosa, 40
Betulaceae, 39
Big Tree, 8
Biota orientalis, 10
Birch, 40, 98
  Common, 41
Black Italian Poplar, 34
Black Mulberry, 48
Black Poplar, 33
Black Walnut, 39
Black Wattle, 66
Blackberry, 47
Bog Myrtle, 37
Bois d'Arc, 48
Bomarea, 29
  caldasiana, 29
  cantabrigiensis, 29
  carderi, 29
  hirtella, 29
  patacocensis, 29
Bo-tree, 50
Bow wood, 48
Bowenia, 2
Box, 76
  Common, 76
Box Elder, 81
Broad-leaved Maple, 81
Bromeliaceae, 23
Buckthorn, 84
  Common, 84
Bull Bay, 53
Bulrush Millet, 15
Buphthalmum speciosum, 105
Burning Bush, 74, 79
Butcher's Broom, 3
Buxaceae, 76
Buxus, 76
  balearica, 76

Buxus—*contd.*
  sempervirens, 76

Cacao, 88
Cactaceae, 92
Caesalpinioideae, 69
Caladium, 23
Californian Laurel, 56
Californian Lilac, 85
Calla, 21
  palustris, 22
Callitris quadrivalvis, 9
Camellia, 89
  Common, 89
  japonica, 89
  reticulata, 89
  sasanqua, 89
  theifera, 90
Camelliaceae, 89
Canoe Birch, 41
Caper, Common, 57
Capparidaceae, 57
Capparis, 57
  spinosa, 57
Caragana, 71
  arborescens, 71
  frutescens, 71
Cardamom Plant, 31
Carica papaya, 55
Caricaceae, 55
Carpinus, 39
  betulus, 39
Carrierea calycina, 92
Castanea, 43
  sativa, 43
Casuarina, 12, 32
  equisetifolia, 32
Casuarinaceae, 32
Caucasian Wing-nut, 38
Ceanothus, 84, 85
  cuneatus, 85
  thyrsiflorus, 85
Cedar of Lebanon, 4
Cedrus, 4
  atlantica, 5
  deodara, 5
  libani, 4

Celastraceae, 79
Celery-leaved Pine, 3
Century Plant, 28
Cephalotaceae, 59
Cephalotus follicularis, 59
Ceratozamia, 2
Cercis, 68
  siliquastrum, 68
Chamaerops, 18
  humilis, 18
Chaste Tree, 101
Cherry Laurel, 53, 65
Chinese Arbor-vitae, 10
Chinese Dwarf Banana, 31
Chinese Persimmon, 98
Chinese Tea Plant, 90
Chorizema ilicifolium, 97
Christ's Thorn, 83
Chrysanthemum, 105
  balsamita, 105
  indicum, 105
  morifolium, 105
  sinense, 105
Cistaceae, 91
Cistus, 91
  creticus, 91
Citronella grass, 15
Cladrastis, 69, 70
  lutea, 70
  tinctoria, 70
Clusia, 90
  flava, 91
  grandiflora, 90
Cluster Pine, 6
Coca Plant, 73
Cocculus carolinus, 53
Coconut Palm, 19
Cocos, 19
  nucifera, 19
Coffea, 103
  arabica, 103
Coffee Plant, 103
Colletia, 85
  cruciata, 85
Colocasia, 22
  antiquorum, 23
Colocasioideae, 23

INDEX

Compositae, 105
Comptonia, 37
  asplenifolia, 37
Copaifera, 67
Coriaria, 77
  myrtifolia, 77
  terminalis, 77
Coriariaceae, 77
Cork Oak, 44, 45
Cornaceae, 94
Cornel, 94
Cornelian Cherry, 94
Cornish Elm, 47
Cornus, 94
  alba, 95
  florida, 94
  mas, 94
  sanguinea, 94
  suecica, 94
Corsican Pine, 7
Corylus, 39
  avellana, 39, 40
  colurna, 40
  maxima, 40
  tubulosa, 40
Costmary, 105
Cotinus, 77
  coggygria, 77
Crack Willow, 35
Creeping Willow, 36
Cucumber Tree, 53, 54
Cucumis asininus, 104
Cucurbitaceae, 104
Cupressus, 10
  lawsoniana, 11
  sempervirens, 4, 10
Cycadaceae, 1, 2
Cycas, 1, 2
  media, 1
  revoluta, 2
Cyperaceae, 16
Cyperus, 16
  esculentus, 17
  longus, 16
  papyrus, 16
  rotundus, 17
Cypress, 4, 10

Dabeocia, 96
  polifolia, 96
Darlingtonia, 58
Date Palm, 17, 18
Deadly Nightshade, 101, 102
Deciduous Cypress, 8
Deodar, 5
Dictamnus, 74
  albus, 74
Dieffenbachia seguine, 22
Dionaea, 59
  muscipula, 59
Dioon, 2
  edule, 2
Diospyros, 97
  kaki, 97, 98
  virginiana, 97
Dogwood, 94
Dorstenia, 48, 49
  contrajerva, 49
Drimys, 55
  winteri, 55
Drosera, 58
Droseraceae, 57, 58
Drosophyllum lusitanicum, 58
Dumb Cane, 22
Durmast, 44
Dutch Elm, 47
Dwarf Birch, 41
Dwarf Cornel, 94
Dwarf Palm, 18

Earth-almond, 17
Ebenaceae, 97
Ecballium elaterium, 104
Edible Pine, 6
Elaeagnus angustifolia, 100
Elaeocarpaceae, 85
Elecampane, 105
Elettaria, 31
  cardamomum, 31
Elm, 45
Encephalartos, 2
  laurentianus, 2
English Elm, 46, 47
Epacridaceae, 97
Ephedra, 12, 13

INDEX

Ephedra—*contd.*
  altissima, 12
  distachya, 13
  nebrodensis, 12
Equisetum, 12
Ericaceae, 96
Erinacea, 70
  pungens, 70
Erythroxylaceae, 73
Erythroxylum, 73
  coca, 73
Escallonia, 60
  illinata, 60
  macrantha, 60
  viscosa, 60
Euberberis, 52
Eucommia ulmoides, 62
Eucommiaceae, 62
Euonymus, 79
  atropurpureus, 79
  europaeus, 79
  japonicus, 80
  latifolius, 79
  verrucosus, 79
Euphorbiaceae, 51
Evergreen Oak, 45
Evergreen Spindle Tree, 80

Fagaceae, 43
Fagus sylvatica, 43
Fan Aloe, 27
Ficus, 49
  bengalensis, 50
  carica, 49
  elastica, 50
  religiosa, 50
  sycomorus, 49
Fig, 49
  Common, 49
Fig Mulberry, 49
Filbert, 40
Flacourtiaceae, 91
Flowering Ash, 99
Flowering Dogwood, 94
Fraxinus, 97
  excelsior, 98
  ornus, 99

Fremontia, 88
  californica, 88
French Willow, 36

Galingale, 16
Garcinia, 90
Genista, 70
  horrida, 70
  radiata, 70
Gentianaceae, 20
Geranium, 77
  robertianum, 79
Ginkgo, 2, 3
  biloba, 2
Ginkgoaceae, 2
Gleditschia, 69
  triacanthos, 69
Glycosmis, 75
  pentaphylla, 75
Glycyrrhiza, 72
  glabra, 72
Golden Wattle, 66
Golden-club, 21
Gnetaceae, 12
Gnetum, 12, 13
  gnemon, 13
Goat Sallow, 35
Gramineae, 15
Great Millet, 15
Grey Alder, 42
Grey Poplar, 32, 33
Guinea corn, 15
Gum Cistus, 91
Guttiferae, 90
Gymnocladus, 69
  canadensis, 69
  dioica, 69

Halimodendron, 71
  argenteum, 71
Hamamelidaceae, 60
Hamamelis, 61
  virginiana, 61
Hazel, 39
Hedgehog Broom, 70
Heliamphora, 58
Helianthemum, 91

INDEX

Helwingia, 95
  rusciflora, 95
Herb of Grace, 74
Hibiscus, 87
  rosa-sinensis, 87
  syriacus, 87
  trionum, 88
Himalayan Blue Pine, 5
Hippocastanaceae, 82
Holm Oak, 45
Honey Locust, 69
Hop-tree, 75
Hornbeam, 39
Horse Chestnut, 82
Huntingdon Elm, 46
Hypericum, 90

Idesia polycarpa, 92
India-rubber Plant, 50
Indian Oleander, 100
Inula, 105
  helenium, 105
Irish Yew, 4
Iron Wood, 61
Italian Alder, 42
Italian Cypress, 10

Japanese Banana, 31
Japanese Laurel, 95
Jateorhiza columba, 52
Judas Tree, 68
Juglandaceae, 37, 38
Juglans, 38
  nigra, 39
  regia, 38
Juniper, 11
  Common, 11
Juniperus, 11
  communis, 11
  drupacea, 12
  excelsa, 12
  sabina, 11, 12
  virginiana, 12

Kaki, 98
Kangaroo Thorn, 66
Kentucky Coffee-tree, 69

Kniphofia, 25
  caulescens, 25
  uvaria, 25
Koelreuteria, 82
  paniculata, 82, 83

Large-leaved Lime, 86
Lauraceae, 56
Laurel Magnolia, 53
Laurus, 56
  nobilis, 56, 65
Leguminosae, 65
Lemnaceae, 23
Lepidobalanos, 44, 45
Leuce, 32
Leucopogon richei, 97
Liliaceae, 25
Lily Tree, 54
Lime, 85
  Common, 86
Liquidambar, 60
  orientale, 60
  styraciflua, 61
Liriodendron, 54
  tulipifera, 54
Locust Acacia, 70
Lombardy Poplar, 33
London Plane, 62
Lycium chinense, 90

Maclura, 48
  pomifera, 48
Macrozamia, 1, 2
  hopei, 1
Madrona Laurel, 96
Magnolia, 53
  acuminata, 53
  conspicua, 54
  delavayi, 53
  grandiflora, 53
  obovata, 54
  stellata, 54
Magnoliaceae, 53
Mahonia, 52
Maidenhair Tree, 2, 3
Malvaceae, 87
Mandragora, 102

INDEX

Mandragora—*contd.*
  officinarum, 102
Mandrake, 102
Manna Ash, 99
Maple, 80
  Common, 80
Maritime Pine, 6
Marsdenia, 100
  erecta, 100
Marsh-mallow, 87
Menispermaceae, 52, 53
Menispermum canadense, 53
Microcycas, 2
Mimosa, 66
Monstera, 20
  deliciosa, 20
Moonseed, 53
Moonseed Family, 53
Moraceae, 47, 51
Morus, 47
  alba, 48
  nigra, 48
Mountain Elm, 46
Mountain Magnolia, 53
Mulberry, 47
Musa, 30
  basjoo, 31
  cavendishii, 31
  ensete, 30
  sapientum, 30
  textilis, 30
Musaceae, 29
Musk Rose, 63
Myrica, 37
  cerifera, 37
  gale, 37
  nagi, 37
Myricaceae, 37
Myrmecodia, 104
  echinata, 104
Myrtaceae, 93
Myrtle, 93
  Common, 93
Myrtus, 93
  communis, 93

Nelumbium, 51

Nelumbium—*contd.*
  lutea, 51
  speciosum, 51
Neosia, 6
Nepenthaceae, 57, 58
Nepenthes, 58, 59
Nerium, 100
  odorum, 100
  oleander, 100
Norway Maple, 81
Nut Palm, 1
Nuttallia, 65
  cerasiformis, 65
Nymphaeaceae, 51

Oak, 44, 98
  Common, 44, 45
Old Man's Beard, 24
Olea, 99
  europaea, 99
  europaea *var.* sativa, 100
Oleaceae, 98
Oleander, 100
  Common, 100
Oleaster, 100
Olive Tree, 99
One-needled Pine, 6
Opuntia, 92
  cantabrigiensis, 93
  ficus-indica, 92
  tuna, 92
Oregon Grape, 52
Oriental Plane, 62
Orontium, 21
  aquaticum, 21
Oryza, 16
  sativa, 16
Osage Orange, 48
Osier, Common, 36
Osmaronia, 65
Oso Berry, 65
Oxalis, 77

Paliurus, 83
  australis, 83
Palm, 35
Palmae, 17

INDEX

Papaw, 55
Paper Birch, 41
Paper Reed, 16
Papilionatae, 69, 70
Parrotia, 61
  persica, 61
Pavia, 82
Pea Tree, 71
Pear, 63
Peltandra virginica, 22
Pennisetum, 15
  americanum, 15
  typhoideum, 15
Persimmon Tree, 97
Phellodendron, 74
  amurense, 74
  japonicum, 74
Phillyrea, 84, 99
  angustifolia, 99
  decora, 99
  latifolia, 99
  media, 99
Philodendroideae, 22
Philodendron, 22
  erubescens, 22
Phoenix, 17
  dactylifera, 17
  sylvestris, 17, 18
Phyllocladus, 3
  trichomanoides, 3
Phyllodineae, 3
Pimenta, 93
  officinalis, 93
Pimento, 93
Pinaceae, 3, 4
Pine, 5
Pineapple, 25
Pinus, 5
  excelsa, 5
  gerardiana, 6
  halepensis, 6
  laricio, 7
  laricio *var*. nigricans, 7
  laricio *var*. pallasiana, 7
  monophylla, 6
  pinaster, 6
  pinea, 11

Pinus—*contd.*
  sylvestris, 7
  sylvestris *var*. scotica, 7
Pistia, 23
  stratiotes, 23
Plane, 62, 70
Plantain, 30
Platanaceae, 62
Platanus, 6, 62, 70
  acerifolia, 62
  occidentalis, 63
  orientalis, 62, 63
Podalyrieae, 70
Podophyllum peltatum, 102
Poison Ivy, 78
Polyalthia longifolia, 5
Poncirus trifoliata, 75
Poplar, 32
Populus, 32
  alba, 32, 33
  canescens, 32, 33
  deltoidea, 34
  italica, 33, 34
  nigra, 33, 34
  serotina, 34
  tremula, 32, 33
  tremula *var*. glabra, 33
  tremula *var*. sericea, 33
Portugal Laurel, 65
Prickly Ash, 74
Prunus, 63, 64, 65
  amygdalus, 64
  armeniaca, 63
  laurocerasus, 53, 65
  lusitanica, 65
  mahaleb, 64
  triloba, 64
Ptelea, 75
  trifoliata, 75
Pterocarya, 37, 38
  caucasica, 38
  fraxinifolia, 37, 38
Purple Osier, 36
Puya, 24
  chilensis, 25
Pyrus, 63
  salicifolia, 63

INDEX

Quercus, 44
  cerris, 44
  ilex, 45
  lanceolata, 45
  obtusata, 45
  pedunculata, 44
  robur, 44, 45
  sessiliflora, 43, 44, 96, 98
  suber, 44, 45

Red Cedar, 12
Red-hot Poker, 25
Redwood, 7, 8
Retama raetam, 11
Retanilla, 85
Rhamnaceae, 83
Rhamnus, 84
  alaternus, 84
  catharticus, 84
  frangula, 84
  purshianus, 84
Rhipsalis, 92
  cassytha, 92
Rhus, 78
  coriaria, 78
  cotinus, 77
  toxicodendron, 78
  typhina, 78
Rice, 16
Robinia, 70
  hispida, 71
  pseudacacia, 70
  viscosa, 71
Rosa, 63
  macrophylla, 63
  moschata, 63
Rosaceae, 63
Rose, 63
Rose-Acacia, 71
Rubiaceae, 103, 104
Rubus, 47
Rue, Common, 74
Ruscus, 3
Rush-nut, 17
Ruta, 74
  graveolens, 74
Rutaceae, 73

Sacred Lotus, 51
St Dabeoc's Heath, 96
St John's Wort, 90
St Lucie Cherry, 64
Salicaceae, 32
Salix, 34
  alba, 35
  alba *var.* caerulea, 35
  alba *var.* vitellina, 35
  babylonica, 35
  babylonica *var.* annularis, 35
  caprea, 35, 36
  cinerea, 36
  fragilis, 34, 35
  purpurea, 34, 36
  repens, 36
  triandra, 36
  viminalis, 36
Sallow, 36
Salt Tree, 71
Sapindaceae, 82
Sapotaceae, 51
Sarracenia, 57, 58
Sarraceniaceae, 57, 58
Savin, 11
Saxifragaceae, 60
Sciadopitys, 7
  verticillata, 7
Scirpus maritimus *var.* umbellatus, 16
Scotch Elm, 46
Scotch Pine, 7
Scrophulariaceae, 103
Semele, 3
Sequoia, 7, 42
  gigantea, 8
  sempervirens, 7, 8
Sessile-fruited Oak, 44
Setaria italica, 16
Shrubby Trefoil, 75
Silver Wattle, 66
Simarubaceae, 75
Skunk Cabbage, 21
Slippery Elm, 88
Small-leaved Elm, 47
Small-leaved Lime, 86

INDEX

Small-leaved Maple, 80
Smooth-leaved Elm, 46, 47
Smyrnium perfoliatum, 87
Solanaceae, 90, 101
Solanum, 102
Sophora, 70
Sophoreae, 70
Sorghum, 15
   vulgare, 15
Spanish Chestnut, 43
Spanish Liquorice, 72
Spanish Moss, 24
Speedwell, 103
Spindle Tree, 79
Squirting Cucumber, 104
Staghorn Sumach, 78
Stangeria, 2
   paradoxa, 2
Sterculiaceae, 88
Steudnera, 23
Stone Pine, 11
Strawberry Tree, 96
   Common, 96
Strobus, 5
Styphelia, 97
   richei, 97
Suber, 45
Sugar Maple, 80
Sumach, 78
Swamp Cypress, 8
Sweet Chestnut, 43
Sweet Fern, 37
Sweet Flag, 20
Sweet Gale, 37
Swertia chirata, 20
Sycamore, 49, 80
Sylvestres, 5
Symplocarpus, 21
   foetidus, 21

Tanekaha, 3
Taxaceae, 3, 4
Taxodium, 8
   distichum, 8
Taxus, 3
   baccata, 4
   fastigiata, 4

Tea Plant, 90
Tetraclinis, 9
   articulata, 9
Thea, 89
Theobroma, 88
   cacao, 88
Thuya, 9
   occidentalis, 9
   orientalis, 10
   plicata, 10
Thyine Wood, 9
Thymelaeaceae, 26, 100
Tilia, 85
   americana, 87
   argentea, 87
   cordata, 86
   europaea, 86
   parvifolia, 86
   petiolaris, 87
   platyphyllos, 86
   tomentosa, 87
   vulgaris, 86
Tiliaceae, 85
Tillandsia, 24
   usneoides, 24
Toothache Tree, 74
Torch Lily, 25
Trachycarpus, 18
   excelsa, 18
Tree of Heaven, 76
Trifoliate Orange, 75
Tsung Palm, 18
Tulip Tree, 54
Tumboa, 12
   bainesii, 13
Turkey Oak, 44
Turkish Hazel, 40
Typha, 78

Ulmaceae, 45
Ulmus, 45
   campestris, 46
   glabra, 46, 47
   hollandica, 47
   major, 47
   minor, 47
   montana, 46

INDEX

Ulmus—*contd.*
  nitens, 46, 47
  sativa, 47
  stricta, 47
  vegeta, 46, 47
Umbellularia, 56
  californica, 56
Umbrella Pine, 7
Utricularia, 24

Venetian Sumach, 77
Venus's Fly-trap, 59
Verbenaceae, 101
Veronica, 103
  cupressoides, 103
Vinegar Tree, 78
Vitex, 101
  agnus-castus, 101

Wafer Ash, 75
Wahoo, 79
Walnut, 38
Warty Spindle Tree, 79
Water Arum, 22
Water Lettuce, 23
Wax Myrtle, 37
Weeping Willow, 35
Wellingtonia, 8
Welwitschia, 12, 13
  mirabilis, 13
White Birch, 40
White Lime, 87
White Mulberry, 48
White Poplar, 32
White Willow, 35

Wig-tree, 77, 78
Willow, 34
Willow-leaved Pear, 63
Winter's Bark, 55
Witch-hazel, 61
Withy, 35
Wych Elm, 46, 47

Xanthoceras, 83
  sorbifolia, 83
Xanthosoma, 23
Xanthoxylum, 73
  alatum, 73
  fraxineum, 74
  planispermum, 73

Yellow Poplar, 54
Yellow-wood, 70
Yew, 3
  Common, 4
Yucca, 27
  angustifolia, 28
  australis, 28
  elephantipes, 28
  filifera, 28
  glauca, 28
  gloriosa, 9, 28
  guatemalensis, 28
  recurvifolia, 28
Yulan, 54

Zamia, 2
Zantedeschia aethiopica, 22
Zingiberaceae, 31
Zizyphus Spina-Christi, 84

www.ingramcontent.com/pod-product-compliance
Ingram Content Group UK Ltd.
Pitfield, Milton Keynes, MK11 3LW, UK
UKHW040656180125
453697UK00010B/206